PHYSICAL ASPECTS OF NATURAL CATASTROPHES

PHYSICAL ASPECTS OF NATURAL CATASTROPHES

by

Adrian E. Scheidegger

Professor,
Institute of Geophysics,
Vienna Technical University, Vienna, Austria

ELSEVIER SCIENTIFIC PUBLISHING COMPANY
AMSTERDAM – OXFORD – NEW YORK 1975

ELSEVIER SCIENTIFIC PUBLISHING COMPANY
335 Jan van Galenstraat
P.O. Box 211, Amsterdam, The Netherlands

AMERICAN ELSEVIER PUBLISHING COMPANY, INC.
52 Vanderbilt Avenue
New York, New York 10017

Library of Congress Card Number: 74-10263

ISBN 0-444-41216-6

Printed in The Netherlands

To
James Tuzo Wilson

PREFACE

The present book aims at a compilation of the physical processes that are involved in the occurrence of natural catastrophes. In this, it straddles the border line between pure science and engineering. With the increase of the number of human beings in the world, more and more areas are being settled which are subject to periodic ravagings by such scourges as earthquakes and tornadoes. Technological advances induce man to undertake projects in regions which may be exposed more and more to the raw forces of nature. Moreover, man himself may change the morphology of the earth's surface so that a latent catastrophe may actually take place: for example, the landslide at Vajont may well have been induced by filling up of the lake in the valley below.

The engineer may be asked to evaluate the risks involved in a certain project and to take effective measures for the protection of lives and property in a catastrophic eventuality. To do this intelligently, the very basis for the design is often missing or hard to obtain: this is an adequate knowledge of the mechanics of the feared phenomena. It is indeed not the engineer's task to study natural phenomena as such; this is the task of the scientist. The scientist, however, is generally not concerned with the phenomena that might be involved in catastrophes, as they may appear to be minor on a global scale and are thus not particularly interesting to him. It is the intention of the present book to attempt to close this obvious gap in our knowledge.

Therefore, the book is essentially a study in natural science. The subject matter, however, was chosen from engineering interests. An effort has been made to provide sufficient quantitative estimates regarding the forces, energy, frequencies, etc., of the phenomena involved, so that an engineer can build design criteria upon it – at least as much as this is possible at the present time, for our detailed knowledge of these phenomena is often still woefully inadequate.

Some of the material presented in this book has been taken from the writer's own articles. Accordingly, he is grateful to the publishers of the *Journal of Hydrology*, of *Géotechnique* (the Council of the Institution of Civil Engineers, London), of the *Proceedings of the Cosenza Conference in 1971 on Natural Slope Stability*, and of *Rock Mechanics* for permission to re-use that material. The Seismological Society of America has given permission to reproduce some figures from its *Bulletin* and the Zentralanstalt für Meteorologie und Geodynamik in Vienna has allowed the reproduction of drawings from its various publications; the writer wishes to thank them.

This book is dedicated to Dr. James Tuzo Wilson, Principal of Erindale College and Professor at the University of Toronto who encouraged the writer to begin his geophysical career. Without Dr. Wilson's stimulating enthusiasm the writer, while a graduate student at the University of Toronto, would never have become introduced to the fascinating subjects of the earth sciences. The dedication is intended to give expression to the writer's deep gratitude for the continuous interest on the part of Dr. Wilson in his scientific development and for a long and happy professional and personal association with him.

A.E.S.
Vienna, November 1973

CONTENTS

Chapter 1. INTRODUCTION

1.1. The scope of the term "catastrophe"

Man tends to term any change in his environment that affects his works or his life in an unwanted fashion as a "catastrophe". Therefore, catastrophes may not only be represented by sudden violent events, such as earthquakes or volcanic eruptions, but also by the gradual slow drowning of a city like Venice – a process that may take centuries to run its course. In this book we shall deal only with (more or less) natural catastrophes: the wilful destruction of a city by a hydrogen bomb will be left out of consideration. Nevertheless, events triggered unintentionally by artificial changes in the environment *will* be considered: the type of event we have in mind is, for instance, represented by the Vajont slide which was (apparently) triggered by the filling up of an artificial lake below a barely stable slope.

This book is a study of the *physical* aspects of natural catastrophes, with a view, of course, to understanding the cause and the course of such events. It is hoped that lessons can be drawn for the future from such a study. The branch of physics involved in the explanation of catastrophes is mostly *theoretical mechanics*. In some instances (lightning, tornadoes, hurricanes) *electrodynamics* and *thermodynamics* also play a role.

Unfortunately, as often happens in modern physics, it is difficult to find in many cases a direct cause-and-effect relationship. Thus, the time and magnitude of an impending or feared catastrophe can seldom be accurately predicted: all one can arrive at is an estimate of the *probability* of its occurrence. Therefore, there will be much discussion in this book of statistical problems, of risk estimates and of danger ratings.

The types of catastrophes man is faced with may originate within the earth (earthquakes, volcanic eruptions); they may originate upon the earth (slopes may become unstable due to internal or external causes); or they may be caused by water in frozen (snow and ice avalanches) or liquid form (floods, coastal effects). Finally they may have their origin in the atmosphere (violent storms). It is the aim of the present study to discuss these various possibilities in detail.

1.2. The basic physical problems

1.21. General remarks

As noted in Sect. 1.1, the basic physical problems encountered in a study of catastrophes belong to theoretical mechanics, thermodynamics and electrodynamics. It is beyond the scope of this book to present the basic aspects of these disciplines; the reader is referred to the many excellent textbooks on these subjects for the necessary background.

Nevertheless, we shall mention here the basic philosophical aspects of the course of a catastrophe: evidently, if a prevailing status quo is preserved, no catastrophe occurs. Therefore, a catastrophe generally entails the *termination* of a stable state: at some time and place, a previous state of stability breaks down. Therefore, the first problem in the analysis of a catastrophe is usually the analysis of the conditions necessary for stability and an attempt at an understanding why these conditions no longer continued.

Once the stability of a system has broken down, events begin to happen. In many cases, the catastrophe is connected with the motion of large masses of material (soil, rock, water, ice), usually downhill in the gravity field. The course of the mass movement often determines the severity of the catastrophe.

Some catastrophes have their origin in the fluid envelopes of the earth: vortices in an unstable atmosphere lead to "severe weather"; "tidal" waves in the sea may impinge upon a coast. The physical background for events originating in the fluid envelopes must be sought in that branch of mechanics which is called hydrodynamics, with additional research in the branches of thermodynamics and electrodynamics. However, even in such systems, it is the breakdown of a "normal" (stable) state in a general sense (vortex formation) which triggers future events. Therefore, one is still faced with the sequence of triggering-effect and subsequent evolution of the events, as this is the normal pattern of the occurrence of a catastrophe.

1.22. Stability of a system

The stability of a system can be of two types: static or dynamic. Engineers have widely studied the conditions for stability – man is mostly interested in maintaining stability, not in knowing how conditions change once stability breaks down. Therefore, many criteria have been established for determining whether a system is stable or not: yield strengths have been found for solid materials, critical numbers have been established for stationary flow, etc. These criteria involve external parameters, and generally tend to *assure* the safety of a system if they are satisfied. They are therefore usually taken on the "safe" side, as they are used for design purposes.

Thus, the violation of a stability criterion does not necessarily mean that the stability of a system will break down; only the converse is true. It is, therefore, not often possible to infer accurately from an estimate of the evolution of the external parameters in connection with an established stability criterion that a stable system will collapse at a certain point. This is one of the reasons why the prediction of disasters is so difficult.

1.23. The motion of mass

One of the possible consequences of the breakdown of the stability of a system is that masses are set in motion. Whilst many criteria are found in the literature for estimating the stability of a system, much less is available on the actual evolution of the events after the stability has broken down: the main concern of mankind is to prevent instability altogether (hence the criteria), not to follow the events thereafter. Nevertheless, the prediction of the severity of an impending catastrophe often depends on a knowledge of its exact course: how far is a landslide going to reach; how fast is a flood wave going to flow down a valley?

The downhill motion of a mass m can be described phenomenologically by a frictional model. Assuming the slope of the "bed" over which the mass slides to be β, one has:

$$m\ddot{x} = mg \sin \beta - F \tag{1.23-1}$$

where x is the coordinate along the bed, g the gravity acceleration, and F a frictional force (see Fig. 1). For "dry" friction, the frictional force is proportional to the normal pressure:

$$F = f\, mg \cos \beta \tag{1.23-2}$$

where f is the coefficient of friction. For viscous friction, F is taken proportional to the velocity $\dot{x}$, and for turbulent friction it is taken proportional to $\dot{x}^2$. However, it is often convenient to characterize an observed motion at least instantaneously by a frictional coefficient f, although the dry-friction law may not actually be valid. The coefficient f is then a variable and is calculated from the observed values of the acceleration $\ddot{x}$. One has:

$$\ddot{x} = g \sin \beta - fg \cos \beta \tag{1.23-3}$$

It is then revealed that the coefficients f for sliding (rock, soil) masses are equal to those that one would expect from laboratory experiments only for small volumes; for large volumes (above

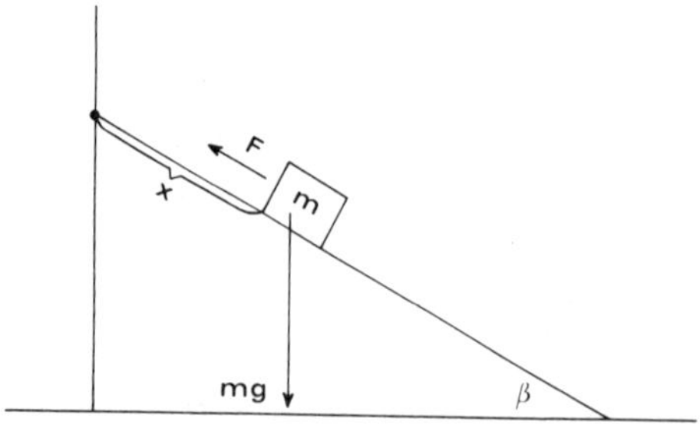

Fig. 1. Definition sketch for a mass moving downhill.

50,000 m^3) they become smaller (cf. Sect. 4.74). The friction is *not* velocity-dependent (it can thus be accurately described by eq. 1.23-3), but decreases with the *volume* of the sliding mass.

The simple frictional model has also been used successfully for describing the motion of soil-rock-water (or air) suspensions. The "land slides" are but one extreme case of this type of motion, since there is the suspicion, on account of the observed low coefficients of friction, that an air cushion or some such feature is formed at the sliding surface. "Slides" with a lower content of solids are represented by fast mud flows (Sect. 4.83), by wet-snow avalanches (Sect. 5.33), by volcanic ash flows (Sect. 5.33) and by turbidity currents in the sea (Sect. 6.52). There appears, in fact, to be a gradual transition between such phenomena.

1.24. Hydrodynamics

The catastrophes due to hydrodynamic effects are mostly caused by *vortices*. An unstable atmosphere expresses itself in the formation of eddies of lesser or greater dimensions where the wind speeds can assume large values. Secondary effects include the separation of electrical charges leading to the phenomenon of lightning and thunder.

Another hydrodynamic phenomenon is the origination of high waves: either flood waves proceeding down a valley or giant waves at sea (usually triggered by an earthquake or a hurricane) impinging upon a coast. These phenomena are governed by the nonlinear gravity wave equations of liquids with a free surface.

There are many treatises on hydrodynamics in which the basic physics underlying the above phenomena have been discussed; the reader is referred to these. The stability conditions for the atmosphere will be discussed briefly in this book in Sect. 7.12. Further references dealing with the problems in question will be given in their proper context.

1.3. The basic statistical problems

1.31. The problem of prediction

The prediction of the time and place of a catastrophe's occurrence is one of the most desirable aims of a study of such events.

The first way to attempt such a prediction is by monitoring indicative precursors. It has been found that specific geodetic changes precede the advent of certain types of volcanic eruptions. There is some hope that eventually similar indicators will be detected which precede the onset of an earthquake. Analogous signs can be found which point to the imminent occurrence of a tornado or a hurricane.

Unfortunately, unequivocal indicators for an imminent catastrophe cannot always be found. Moreover, at best such indicators exist just before the occurrence of a catastrophe; long-range forecasts cannot generally be made. Nevertheless, it would be desirable to have some idea of the frequency and severity of catastrophes that may be expected to occur in a certain area. Such information can only be obtained by the application of statistics. A direct evaluation of observational data can give one an impression of the past frequency of the occurrence of catastrophic events. Under the assumption that the probabilities do not change with time, the same frequencies that were observed in the past are predicted for the future. However, the data usually refer only to a limited time interval. If no flood or earthquake above a certain intensity was observed during the past 50 years, the question arises whether it is probable that a more intense event might occur during the next 200 years. Thus, the problem is one of extrapolating the available data to the future. Such an extrapolation must be based upon the assumption of a statistical model – which is always to a certain extent arbitrary. Examples of models will be found throughout this book.

1.32. Statistics of rare events

Many types of catastrophes are of sufficiently short duration to be considered as instantaneous events in time. Therefore, the basic observational data of catastrophes consist of a list of the times T_i

of their occurrence (for example, a list of dates on which earthquakes, floods, tornadoes, etc. occurred at a certain locality). Very often, a parameter Q_i indicating the intensity of the i-th catastrophe can be defined. Thus, the raw material from the observations contains not only a list of times T_i, but also corresponding values of the intensity Q_i. From this set of data statistical predictions are to be attempted.

In order to analyse the raw data, it is first of all necessary to classify the catastrophes according to their strengths Q_i. New lists are drawn up which contain only those events where Q_i was greater or equal to some given value Q_e of Q_i. These are lists of the Q_e-events. The statistical properties of the Q_e-events are then studied. These will be different for the various values of Q_e.

The *first way* to analyse the statistical properties inherent in a list of Q_e-events is to focus one's attention on the frequency of such events. A reference time interval of unit length is chosen (this may be a year, a decennium, a millennium, etc.) and the number n of times a Q_e-event occurred in any one unit interval, is determined. In this fashion one arrives at an empirical frequency distribution function $f(n)$.

The *second way* for a statistical analysis of the raw data concerning Q_e-catastrophes is by focussing one's attention on the time *intervals* t_i between subsequent Q_e-events. One thus arrives at a stochastic series of numbers t_i representing the periods of return of the catastrophes whose statistical properties can be studied. One can again introduce a probability function $F(t)$ giving the probability that a return period has not ended up to the time t. It is possible to construct empirical curves for $F(t)$ for given types of catastrophes.

In many instances, the catastrophes appear to be produced by a Poisson process. For such a process, the probability $p(j,t)$ of j events occurring in the time interval t is given by:

$$p(j,t) = \frac{(kt)^j}{j!} e^{-kt} \qquad (1.32\text{-}1)$$

The recurrence time, then, is exponentially distributed (note that $F(t) = p(1,t)$) and the mean and variance of the number of events per unit time are both equal to k [note that $f(n) = p(n,1)$].

The Poisson process is completely determined by the single parameter k. Therefore, it is frequently used as a simple stochastic model of catastrophes.

The discussion above refers only to a specific level Q_e of the intensity of the catastrophes; of greatest importance is the problem of finding the dependence of $f(n)$ or $F(t)$ on Q_e. For a Poisson distribution, one only needs to find the dependence of k on the intensity:

$$k = k(Q_e) \tag{1.32-2}$$

First of all, this is done empirically on the basis of the available data, which leads to an observed relationship between mean frequency and intensity. If a simple analytical form is found, it is then hopefully extrapolated to values of Q_e which have not yet been observed. Often, attempts are made to estimate a maximum possible Q_e from other indications beyond which the $k(Q_e)$ relationship is not to be extrapolated. In this fashion, a complete stochastic model of the process producing the catastrophes is obtained.

1.33. The estimation of risk

The statistical procedure indicated in Sect. 1.32 leads to an estimation of the risk a locality will undergo with regard to some possible catastrophes. The relationship $k(Q_e)$ gives the estimated mean frequency of a catastrophe of intensity Q_e in unit time. This can also be expressed in terms of mean recurrence intervals t_e for catastrophes of strength Q_e. Assuming the catastrophes to be caused by a Poisson process (which is not always justified), one can calculate the standard deviations from the mean, etc. Thus, the curve $k(Q_e)$ [or $t_e(Q_e)$] represents the fundamental estimate of the risk of a locality with regard to the type of events under consideration.

As is shown in Sect. 1.32, the nature of the extrapolation procedure for arriving at the $k(Q_e)$ curve is always to some extent arbitrary. This should be kept in mind when risk estimates are used for practical applications.

1.4. Units

In any multidisciplinary study, the problem of units to be used presents itself. Each discipline has its customary measurement units, which often differ from those used by a related discipline. In our study we are chiefly concerned with engineering applications of "pure" science. In engineering, the system of units is generally based on a unit of *force*, time and length. On the other hand, in science the system of units is based on a unit of *mass*, time and length. Thus, one is faced with a basic dichotomy, inasmuch as the gravity acceleration *g* (which differs in various regions on the earth) enters into any conversion formulas. It would appear to be futile to try to convert all units into one consistent system in this book, inasmuch as all feeling for orders of magnitude would then be lost.

However, we shall use exclusively *metric* units in our study, i.e., the English units of foot (mile) and pound (mass or force) will be converted to the corresponding metric units. However, the metric system is by no means consistent. In science, the preferred units are cm, g (mass) and sec; in engineering, m (or cm), kp (kilogram force) and sec. Some European countries have made efforts to introduce yet another system, viz., one based on a consistent application of m, kg (mass) and sec as basic units. However, this writer feels that it only increases confusion, and that it may be some time yet until atmospheric pressures will generally be given in Newton m^{-2} (the m-kg-sec unit of pressure), distances between continents in m (not km) and the power of an automobile in Watts (not m kp sec^{-1}).

Therefore, in this book two basically different systems of units will be used side by side: the scientific system based on cm, g (mass) and sec, and the engineering system based on m, kp and sec. Simple powers of ten of these units will be used: km (not m or cm) and pressures in kp cm^{-2} (not kp m^{-2}) or in bars (1 bar = 10^6 dynes cm^{-2} = 10^3 millibar). Furthermore, for slow movements, the time units of "day" and "year" (sea-level changes in mm $year^{-1}$ and not in m sec^{-1}) have been used so as to enable the reader to acquire a feeling for the orders of magnitude involved. Although this procedure introduces a certain number of inconsistencies, it is believed to be preferable to an exclusive application of one of the systems which would lead, in part, to entirely unfamiliar units.

Chapter 2. EARTHQUAKES

2.1. Introduction

Earthquakes have caused some of the most violent catastrophes mankind has known. Physically, they represent shocks originating at or near (to a depth of some 600 km) the earth's surface, and they may be so violent that buildings collapse and huge "tidal" waves (tsunamis) are created. The area most affected by an earthquake is called its epicentral zone; however, the effects of any large earthquake may be felt, or can at least be measured, all over the world.

For the registration of earthquakes, one uses seismographs, instruments designed to respond to and register the motion of the ground they stand on. The seismograph records a seismogram; this is a line which is related to the motion of the earth in any one chosen direction. Any change of frequency in the seismogram of an earthquake is called a *phase*. The principal phases in the seismogram of an earthquake have been called P, S and L phases; it has been established that they represent the onset of compressional, transverse bodily and surface waves, respectively. Seismographs may be constructed to respond primarily to certain frequency intervals, or to strong motion only. Indeed, the difference in amplitude in the epicentral zone of a large earthquake and its registration half-way around the world is so great that it is impossible to have one set of instruments suitable for all purposes.

This is not a treatise on the science of earthquakes ("seismology"). For general information, the reader is referred to the various excellent monographs on the various aspects of the subject, such as those by Savarenskii and Kirnos (1955), Richter (1958) and Bullen (1965). Here we shall be concerned solely with the physics of the earthquake source and the effects of an earthquake on the ground. The layout of this chapter is as follows. First (2.2) the earthquake source itself will be described, then (2.3) the effect of earthquakes on the ground will be presented; following this, we shall discuss the geographical (2.4) and temporal (2.5) distributions of earthquakes and, finally (2.6), we shall deal with the problems of seismic risk, seismic zoning and earthquake prediction.

2.2. The earthquake source

2.21. General description

As noted above, earthquakes commonly appear as "shocks". This term implies that the event has a specific "time instant", at which it "occurs". In fact, this is an oversimplification, inasmuch as it is hardly conceivable that the "shock" occupies only an infinitesimal time interval. Actually, the motion does, of course, take time – it may even be quite slow as borne out by the repeated bradyseisms ("slow earthquakes") in the Phlegrean Fields near Naples in which vertical motions of up to 7 m in 48 h were observed (Oliveri del Castillo and Quagliarello, 1969). This type of phenomenon, however, is rather exceptional; one is generally faced with conditions in which the earthquake can be approximated by a "shock" which has at least a well-defined temporal beginning. The time-instant of its beginning at the source is the "origin time" (in Universal Time) of the earthquake.

In a similar fashion, one also usually assumes that the earthquake source has a well-defined location in the earth which can be specified by giving its *geographical coordinates* ("epicentre") and its *depth* (in km) below the earth's surface. This is an even greater oversimplification than the statement of an origin time, but one can again take recourse to the assumption that one is concerned with the location of the source ("focus") where the motion *begins*.

Therefore in the common treatment of earthquakes one assumes that the motion has a specific beginning and a definite origin. Its effect on the ground, of course, is known to last for some time.

2.22. *Causes of earthquakes*

Regarding the mechanical causes of earthquakes, for a long time various types have been distinguished. In fact, earthquakes are classified accordingly into several classes: (a) the collapse of a cavity and similar effects; (b) volcanic shocks; and (c) tectonic effects.

At one time the collapse of cavities was thought to be *the* major cause of earthquakes (Volger, 1858). However, it is now clearly apparent that this cause can be the origin of minor tremors only.

Volcanic eruptions, on the other hand, can cause far more serious shocks: there is no doubt that volcanic activity (Schick and Ruiscetti, 1973) can cause tremors, and volcanic explosions like that of Krakatoa (Rittmann, 1960) or Thera (Santorin) (Marinatos and Imboden, 1973) have caused large earthquakes.

The most common "cause" of earthquakes is, however, tectonic activity (Trifunac, 1972; Schick, 1972). It has been proven that (nonhydrostatic) stresses are present everywhere in the earth's crust and upper mantle; in many areas, the stress state is such that the material is on the verge of instability. Earthquakes are thus closely related to faulting; in fact, the mechanism of an earthquake focus may be naively represented as a fault. This faulting mechanism can be shown to fit into the local orientation of the global stress system. If the material is close to breaking point, then, quite reasonably, relatively minor causes could "trigger" earthquakes. In this instance, Sytinskiy (1963) and Simpson (1968) claim to have found that solar activity plays a significant role in the triggering of earthquakes. However, such claims are still based on tenuous statistical correlation analyses.

More convincing are the correlations between earthquakes and hydrological phenomena. These will be discussed in Chapter 6, Sect. 6.28.

2.23. *Seismic source models*

As noted above, it is usually possible to consider an earthquake source as represented by a focus that has a specific location and becomes activated at a particular time. Then there is the question of the physical processes that may occur at such a focus. These physical processes must produce both the seismic wave patterns noted at observatories and the effects experienced in the epicentral region.

Understanding the physics of the seismic source is not easy. The general implications of this problem have been discussed by various writers (Chinnery, 1960; Stauder, 1962; Scheidegger, 1963b; Benioff, 1964; Aki, 1972a, b). Accordingly, the attempts at understanding the seismic source mainly rely on mathematical and physical models. These models may be "naive" ones, or they may regard the source as a singularity in an otherwise continuous elastostatic or elastodynamic stress field. Attempts have also been made to construct models in order to comprehend the mechanisms of specific earthquakes (e.g. Trifunac, 1972).

The simplest models are the "naive" ones: the earthquake is regarded as a fault in the earth, along which an instantaneous sliding motion is taking place. This immediately explains the observed patterns of the directions of the onset of P-waves in the seismograms. Conversely, a compilation of the seismograms from records of the seismic stations over the world for a particular earthquake enables one to obtain a "fault plane solution" for that earthquake. The technique has been described in detail, e.g. by Scheidegger (1963a).

The naive "faulting" model meets with some serious difficulties. It can easily be shown (Orowan, 1960) that the shear stresses that would be necessary to cause the sliding on the fault surfaces at any but the most shallow depths (to overcome friction) are much greater than the shearing strength of any material that could possibly be imagined to exist at these locations. Consequently, "faulting", if it occurs, must represent some kind of rheological instability rather than an actual fracture (Orowan, 1966; Riecker, 1972; Wu et al., 1972; Molnar et al., 1973; Achenbach and Abo-Zena, 1973) with and without stick-slip. Shear melting (Orowan, 1966), phase transformation processes (Evison, 1967a; Ringwood, 1967; Ritsema, 1970) and the effects of a pore fluid (Riecker, 1966; Robson et al., 1968) have also been advocated as being able to provide the required "lubrication". Evison (1966a, b) has investigated models of an entirely different type, which could not be represented at all as a type of "fault". These models may be applicable to *some* earthquakes, but for *most* there is no question that the "fault" picture is a valid analogy: the innumerable fault plane solutions (Fara, 1964) and the regularity with which these fit into the tectonic stress patterns of the world (Scheidegger, 1965) would otherwise be outright miraculous.

Although the physical processes occurring at an earthquake focus are by no means clear, the "naive" picture of a "fault" must be maintained. The next task is then to give a mathematical description of such a "fault" arising within the earth. Since the regions in which earthquakes occur (to 600 km depth) are commonly regarded as elastic, efforts have been directed to construct mathematical sources in a stress field in an elastic substance that would correspond phenomenologically with a "fault".

The first possibility is represented by elastostatic singularities: two states of the stress field are considered, one before and one after the development of the singularity.

The most commonly used singularity is some sort of singular force. It appears that a dipole force with a moment applied at the focus produces exactly the same nodal lines as those obtained by the model of rigid faulting. Other types of sources have also been tried; it appears that some produce similar nodes as a dipole force with a moment (cf. Scheidegger, 1963a). A distinction can only be made if either S-nodes or amplitudes are taken into account (Byerly and Stauder, 1958). A more complicated force-source, viz. a compensated linear-vector dipole, has been investigated by Knopoff and Randall (1970).

Instead of a singularity in the *stresses* (force), the attention can be directed to singularities in the *strains*. This leads to the dislocation theory of an earthquake focus. Such theories were proposed by Saito (1958), Vvedenskaya (1958, 1959, 1960, 1961), Tesseyre (1961, 1969) and others. Reviews of this theory have been given by Constantinescu (1965) and by Press (1967a). Steketee (1958a, b) has mathematically analysed dislocation models rather closely. He has shown that dislocations can give rise to the same nodal lines as force-models; however, the dislocation model corresponding to a quadrupole is in static equilibrium, but that corresponding to a dipole is not. Furthermore, he noted that in a medium, subject to invariable stresses at its boundaries, the introduction of a dislocation surface would always increase the strain energy so that "ruptures" would be impossible. However, Savage (1969) showed that this is not true if the entire system is considered. The general equivalence of body-force sources and displacement sources was also studied by Nyland (1971).

The theory of dislocations has been extended to volume sources (Scholte, 1962; Scholte and Ritsema, 1962; Braslau and Lieber, 1968), and also to asymmetrical dislocations (Schäffner, 1961). Furthermore, Weertman (1964) and Walsh (1968) investigated the possibility of including friction at the dislocation surface.

The above results are based on elastostatic considerations only. In order to explain the observed seismograms, elastodynamics must be introduced. This can be done by considering the "static" source as multiplied by a harmonic time factor $e^{i\omega t}$, or by actually making models of the initial phases of the transient process of origination of the singularity.

Thus, Knopoff and Gilbert (1960) considered (multipole) force models and made a first-motion approximation. They showed, incidentally, that force sources and displacement sources yield identical first motion radiation patterns in their approximation.

Dynamic force (multipole) models of earthquake sources have been investigated by several researchers. Randall and Knopoff (1970) and Scholte and Ritsema (1962) used a harmonic excitation source, Sato (1969) used a step-function in time and Ben-Menahem (1962) a moving source.

Similarly, dynamic dislocation models were investigated. Archambeau (1968) investigated harmonic sources, Marnyama (1963) and Randall (1964a) step-functions in time, and Vvedenskaya (1965) and Aki (1968a) moving dislocations.

Special elastodynamic solutions have been calculated by Randall to analyse his particular models of a sudden volume change (Randall, 1964b) and a sudden phase transition (Randall, 1966).

The analytical solutions referred to above are at best approximations (harmonic source, first motion approximations, etc.) to the real conditions. Model studies which represented more realistic conditions have also been attempted. Such experiments have been performed, e.g., by Healy and Press (1959, 1960; also Press, 1958) and by Schick and Schneider (1964) using a harmonically excited transducer as seismic source at a "fault". Savage and Mansinha (1963) observed the radiation pattern of elastic waves from a tensile fracture caused by thermal stresses, and Pfluke and Howell (1964) studied the pattern caused by a "locked" fault in plexiglass that was reactivated by external stresses. In general, the theoretical predictions were confirmed.

Special investigations have also been undertaken to collate the theoretical work referred to above with observations from particular earthquakes. Hastie and Savage (1970) constructed a dislocation model for the 1964 Alaska earthquake. Ritsema (1967) tried to fit models to shallow-focus, and Sugimura and Uyeda (1967) to deep-focus earthquake observations.

The results from the "naive" models were generally confirmed by the more complicated analytical and physical models, at least in broad outline. The collation with observational data fits quite reasonably.

2.24. Source parameters

A. General remarks

The various source models discussed in the foregoing section can be characterized by a variety of parameters. The choice and values of these parameters are, of course, significant only if the model to which they refer is specified or at least implicitly understood.

B. Magnitude

Some of the parameters, however, have a more universal meaning, for instance, the total energy liberated by a focus during a seismic event. Unfortunately, it is difficult to determine the energy radiated by an earthquake. Therefore, in relation to focal processes a suitable characterization of the strength of an earthquake had to be determined empirically, at least originally. One such measure is the "magnitude".

The term "magnitude" was defined empirically by Richter (1935). Accordingly, only the maximum amplitude of a seismic record is considered. The relation between the magnitudes M of two earthquakes at a standard epicentral distance of 100 km, and the *maximum* recorder trace amplitude B (mm) is then given by:

$$M_1 - M_2 = \log B_1 - \log B_2 \qquad (2.24\text{-}1)$$

where the seismogram is supposed to have been recorded on a (then) standard torsion seismometer of free period of 0.8 sec, static instrumental magnification of 2,800 and damping ratio of 50 : 1. The zero of the scale is defined by setting M = 3 for B = 1 mm.

As is seen from the above definition, the "magnitude" depends on a particular type of instrument. If other types of instruments are used, the problem arises of linking the various magnitude scales thus obtained. This problem is not uniquely solvable, because such solution would assume that analogous parts of the seismograms of two earthquakes are proportional, which is not true even for identical instruments. The magnitude, thus, cannot be expected to characterize uniquely the "strength" (i.e. energy output) of the focus of an earthquake. Nevertheless, "magnitude" gives a rough

TABLE I

Relations of K to other earthquake parameters (modified after Båth, 1966a)

K	E_0 (ergs)	M	m	$\log V$ (cm^3)	I_0	a_0 (cm/sec^2)
14	10^{14}	1.2	3.3	11.34	0.3	0.4
15	10^{15}	1.9	3.7	12.37	1.3	1
16	10^{16}	2.6	4.1	13.40	2.3	2
17	10^{17}	3.3	4.6	14.43	3.3	4
18	10^{18}	4.0	5.0	15.46	4.3	10
19	10^{19}	4.7	5.5	16.49	5.3	20
20	10^{20}	5.4	5.9	17.52	6.3	40
21	10^{21}	6.1	6.3	18.55	7.3	100
22	10^{22}	6.8	6.8	19.58	8.3	200
23	10^{23}	7.5	7.2	20.60	9.3	400
24	10^{24}	8.2	7.7	21.63	10.3	1000
25	10^{25}	8.9	8.1	22.66	11.3	2000

characterization of a focus and is easily determinable on a seismogram, even if it is not entirely unique for the same earthquake when recorded on instruments in many parts of the world. Båth (1969) compiled the techniques that are used for magnitude determinations by most seismic observatories of the world. Accordingly, a body-wave magnitude m and a surface-wave magnitude M were obtained which coincide at 6.8, but cannot be made to coincide elsewhere (Terashima, 1970; Howell, 1972). Table I shows the relations in question (after Båth, 1966a).

The largest earthquakes reach a magnitude between 8 and 9.

C. Energy

An unambiguous indication of the "strength" of an earthquake would be obtained if it were possible to observe the amount of energy radiated by it. As noted above, however, it is very difficult to do this.

In order to study this problem, one has to calculate the total energy emitted by the earthquake source in elastic waves from the energy flux passing through a seismic station, which is expressed on the trace of the seismograph. Because the radiation pattern of the source enters into such calculations, the specific source model used is also a factor. The calculations involve integrations of the squares of the amplitudes of the seismic recordings (cf. Gutenberg and Richter, 1942; Solovyev, 1959). A good summary of the procedures has been given by Båth (1966a).

The energy is measured in ergs; as energy-class K of an earthquake one has designated the power of 10 of the number of ergs released:

$$K = \log_{10} E \text{ (ergs)} \qquad (2.24\text{-}2)$$

There exists a relation between the energy and various magnitude scales, which is as follows (see King and Knopoff, 1969):

$$\log E = \alpha + \beta M \qquad (2.24\text{-}3)$$

where α and β are empirical coefficients. These can only be determined approximately because it cannot be expected that there is a single relation between the energy (obtained by an integration) and the magnitude (obtained by taking a maximum amplitude) of a wave train. The commonly used empirical relations are shown in Table I. This table was calculated with the values $\alpha = 12.32$, $\beta = 1.42$ in eq. 2.24-3 for the surface-wave magnitude; however, values for α ranging from 6.1 to 13.5 and for β ranging from 1.2 to 2.0 have also been quoted (Båth, 1966a).

D. Strain

In a strained elastic body, the strain is proportional to the square root of the elastic energy stored. The complete relation is:

$$2W = \lambda\theta^2 + 2\mu\epsilon_{ij}^2 \tag{2.24-4}$$

where W is the elastic energy per unit volume, θ is the cubical dilatation, ϵ_{ij} the strain tensor, and λ, μ are the Lamé parameters.

It is therefore possible to interpret the energy released in an earthquake in terms of a corresponding strain release (Benioff, 1955). This procedure, however, presupposes that the earthquake volume is constant for shocks of all magnitudes, which is unlikely to be true. As is evident from eq. 2.24-4, the total elastic strain energy of the focal region is:

$$E = VW \tag{2.24-5}$$

where V is the earthquake volume.

It will be shown below (in Sect.E) that the earthquake volume V increases with magnitude in the same manner as E so that it appears that W, and therewith ϵ, is independent of magnitude (cf. Båth and Duda, 1964).

A quantity related to strain is the deformation D:

$$D = \epsilon V \tag{2.24-6}$$

This, again, increases with magnitude like the volume or the energy, since the strain per unit volume ϵ is constant.

Physically, the independence of magnitude of the strain per unit volume in an earthquake zone signifies that earthquakes of different magnitudes differ by the various sizes of their focal regions. The earthquake occurs when a critical strain in that region is reached.

E. Volume

Another parameter of an earthquake source is its volume. This is that region around the focus in which the elasticity conditions no longer apply. In the models which assume a singularity, this volume would be zero: this is evidently a physical impossibility.

Båth and Duda (1964) identified the earthquake volume with the total volume occupied by the "aftershocks" (cf. Sect.2.53) of an earthquake. By this simple procedure they found that volume increases with magnitude as does the energy:

$$\log V = \alpha + \beta M \qquad (2.24\text{-}7)$$

(but with different coefficients (α = 9.58 and β = 1.47 for the surface wave magnitude) than in eq. 2.24-3). The values obtained, calculated from Båth's and Duda's equation, are shown in Table I.

Eq. 2.24-7, together with eq. 2.24-5 imply that the strain per unit volume in an earthquake focus is constant, as was mentioned above. Alternatively, if one assumes that the strain in an earthquake is always the same, one has a means of calculating (at least relative) volumes (Duda, 1970). These depend on the critical strain assumed. A similar argument was used by Dzhibladze (1971a).

F. Stress

If we consider that the limiting strain in an earthquake area is constant, the same must apply for the limiting stress, since there is a single relation between stress and strain in the elasticity theory: the latter is assumed to be valid until reaching the focal region.

When an earthquake occurs, the stress drops at the (hypothetical) fault surface; this stress drop corresponds to the shearing strength. The stress drop was estimated by Chinnery (1964) on shallow earthquakes by measuring the horizontal displacement at the surface of the fault and assuming that the vertical plane of the fault can be treated as the slip patch which is displaced by a dislocation. In this fashion, stress drops of 10–100 bar were observed. A summary of determinations has been given in a later paper of Chinnery (1969a). The stress drops for earthquakes in the magnitude range 6.8–8.3 are generally between 30 and 60 bars, with extremes from 10 to several hundred bar (Trifunac, 1972). Compared with the usual values for the strength of the upper layers of the earth, this is a very low value. There is no apparent correlation between the "average" stress drop and the magnitude of the accompanying earthquake.

G. Fault length, width and movements

In many shallow earthquakes, one can observe concurrent faulting at the surface. For such shallow earthquakes, relations between the length and the fault L, the (horizontal or vertical) relative movement D and the magnitude M were found (Tocher, 1958; Iida, 1959, 1965; Press, 1967b; Wyss and Brune, 1968; Wyss and Molnar, 1972; Randall, 1972).

In addition, the fault width W and the seismic moment μLDW (where μ is the rigidity) have been introduced (Berckhemer, 1962).

Most of the relations are of the form:

$$M = c_0 + d_0 \log L \tag{2.24-8}$$

$$M = c_1 + d_1 \log D \tag{2.24-9}$$

$$M = c_2 + d_2 \log W \tag{2.24-10}$$

This also implies linear relations between combinations of the "source parameters":

$$M = c_4 + d_4 \log LD \tag{2.24-11}$$

$$M = c_5 + d_5 \log LDW \tag{2.24-12}$$

$$M = c_6 + d_6 \log LW \tag{2.24-13}$$

and so on.

King and Knopoff (1968) used a more complicated relation:

$$M = c_n + d_n \log LD^n \qquad (2.24\text{-}14)$$

where $n \approx 2$. Chinnery (1969b) reviewed the attempts to correlate the above type and found that there is no linear relationship as proposed in eq. 2.24-8 or eq. 2.24-10 that would be valid over the whole magnitude range, but that the linear relationship expressed in eq. 2.24-9 is rather well supported for $3 < M < 8$. A real confirmation of these results is, however, difficult because of the scarcity of good data.

2.3. Effects on the ground

2.31. Phenomenological description

The disastrous effects of earthquakes are caused by the events occurring on the ground near the epicentre: the main phenomenon is a macroscopic motion of the ground which, in a quasioscillatory fashion, may last for some time. This is, of course, the result of the impingement of elastic waves radiating from the earthquake focus on the surface of the earth. It is this motion, and particularly the horizontal component of it, which causes buildings to collapse.

In addition to the shaking motion, discontinuities may arise on the ground. Fissures and, more importantly, lateral and dip-slip faults may form. Large-scale uplifts may also occur; in the earthquake of May 24, 1968 near Inangahua (New Zealand), Lensen and Otway (1971) found that an area of 1,000 km^2 was uplifted by an average of 1 m.

The phenomenological effects mentioned above can be placed on a more quantitative basis as will be shown below.

2.32. Intensity scales; isoseismals

The severity of the effects on the ground due to an earthquake at any point are expressed on an "intensity" scale. The intensity is therefore not a property of the earthquake focus, but represents solely the observed effects on the ground at a particular point. Naturally, "strong" earthquakes (in terms of magnitude or energy) can cause stronger effects than weak ones. The distance from the epicentre, focal mechanism, etc. will, however, have a paramount influence upon the intensity which is actually experienced at a particular point on the earth's surface.

The intensity of an earthquake at various points is determined from inspections and from interviews with witnesses. Rossi (1883) and Forel (1884) developed a 10-point scale, which was extended later by Mercalli (1902) to 12 points following a suggestion of A. Cancani. This scale was further modified by Wood and Neumann (1931) leading to the "modified Mercalli" (M.M.) scale. Various

TABLE II

Shortened version of modified Mercalli scale

Intensity	Effects
I	Not felt by persons; registration on seismographs only.
II	Barely noticeable by favourably placed persons (upper storeys in a house).
III	Weak earthquake; felt by some persons indoors; weak swinging of hanging objects.
IV	Noticed indoors by many, out of doors by a few persons; vibration like from a passing truck.
V	Felt by persons out of doors; sleeping people may be awakened.
VI	Felt by everybody; many people are frightened; some glasses and dishes broken.
VII	Damage to buildings; people have difficulty to remain standing; furniture may be broken; fall of loose plaster, bricks.
VIII	Partial collapse of poorly built structures; chimneys fall.
IX	General panic; large cracks in the ground; ordinarily built houses collapse.
X	General collapse of buildings; large landslides triggered; serious damage to dams, levees etc.
XI	Catastrophic damage; rails bent greatly; underground pipes completely out of service.
XII	Large modification to landscape; large rock masses displaced; nearly total damage to buildings.

minor modifications of the latter are now in common use; the total number of intensity scales is about 50. In America, a version developed by Richter (1958) is used; in Europe, a similar version developed by Medvedev et al. (1967) is commonly applied. The intensity scales are all somewhat problematic (Ambraseys, 1969) because they rely heavily on observations of people and on effects the earthquake has on the construction of buildings. In sparsely populated areas and in areas where the construction of buildings differs from that in Europe or America, it is difficult to determine unequivocal intensity values.

It may be noted that the various 12-point scales differ relatively little from each other. In Table II, we give a shortened version of the gist of these scales. Formally, the 12-point scales can be made to correspond to the 10-point (Rossi-Forel) scale by the following relation:

$$I_{\text{10-point}} = 1.12\, I_{\text{12-point}} - \tfrac{1}{4} \qquad (2.32\text{-}1)$$

In spite of all the difficulties, the intensity reports of an earthquake yield a rough description of the effects on the ground of this earthquake. Such reports are usually represented by means of *isoseismal lines* drawn on a map; these lines separate regions of different intensity values observed in a particular earthquake. As an example, we give in Fig. 2 the isoseismal lines given by Gangl (1969) of the earthquake of Oct. 8, 1927, near Schwadorf in Austria.

The intensity decreases with distance from the epicentre (cf. Peterschmitt, 1952), but by no means in a homogeneous fashion. For the earthquake cited above, the drop in the intensity with distance from the epicentre in various directions is shown in Fig.3 (after Gangl, 1969). There are several formulas purporting to represent this dependence. The first was proposed by Kövesligethy (1900) and Gassmann (1925):

$$I = C - 3 \log \sqrt{r^2 + h^2} - 3\alpha\sqrt{r^2 + h^2} \log e \qquad (2.32\text{-}2)$$

where C is a constant depending on the maximum intensity of the earthquake, r is the epicentral distance, h the focal depth and α

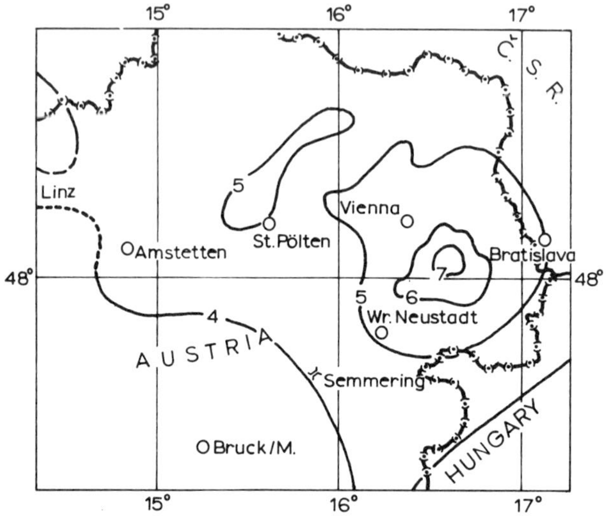

Fig. 2. Isoseismal lines of the earthquake near Schwadorf, Austria, on 8 Oct. 1927. (After Gangl, 1969; reproduced by permission of the Zentralanstalt für Meteorologie und Geodynamik, Wien.)

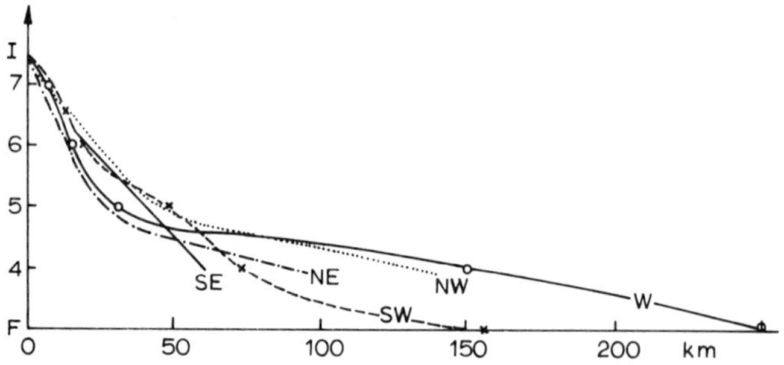

Fig. 3. Drop-off of intensity with distance from the epicenter in various directions for the earthquake of 8 Oct. 1927 near Schwadorf, Austria. (After Gangl, 1969; reproduced by permission of the Zentralanstalt für Meteorologie und Geodynamik, Wien.)

the absorption coefficient. A simpler formula was proposed by Gutenberg and Richter (1942); it is as follows:

$$I = C - 6 \log \sqrt{r^2 + h^2} \qquad (2.32\text{-}3)$$

where the symbols have the same meaning as before.

An important quantity is the maximum (M.M.) intensity that is observed in an earthquake. The maximum intensity depends on the focal parameters of the earthquake, notably on its magnitude M and depth h. Generally, one has an equation of the form (Båth, 1967):

$$I_0 = c_1 M - c_2 \log h + c_3 \qquad (2.32\text{-}4)$$

Gutenberg and Richter (1956) found for shocks in California (h = 16 to 18 km) approximately:

$$I_0 = \tfrac{3}{2} M - \tfrac{3}{2} \qquad (2.32\text{-}5)$$

which they claim to be close to a world average for shallow earthquakes. This equation leads to the values of I_0 given in Table 1 (after Båth, 1966a).

For earthquakes of varying depths h, Shebalin (1968) found in the U.S.S.R. average values for the coefficients in eq. 2.32-4 of $c_1 = 1.5$, $c_2 = 3.5$ and $c_3 = 3.0$. For the mountainous regions of the Caucasus, Dzhibladze (1971b) found $c_1 = 1.5$, $c_2 = 4.2$, $c_3 = 4.0$; whereas Karnik (1965) claimed that generally a more solid relationship would exist when $c_1 = 1.5$, $c_2 = 2.65$, $c_3 = 2.1$. These coefficients depend evidently on the region studied. The uncertainties reflect the ambiguity of the intensity concept.

2.33. Displacement, velocity, acceleration and spectra

As was noted above, the intensity as defined in the last section is a somewhat qualitative concept. Much more important are the numerical values of physically measurable quantities, such as velocity, acceleration, ground amplitudes, etc., in the vicinity of an earthquake epicentre. It has become possible to measure such quantities with the introduction of strong-motion seismographs which are triggered by an occurring earthquake. Strong-motion seismographs are basically designed as accelerometers. The displacement and the velocity of the ground are then found by integration from accelerograms. There are now a large number of

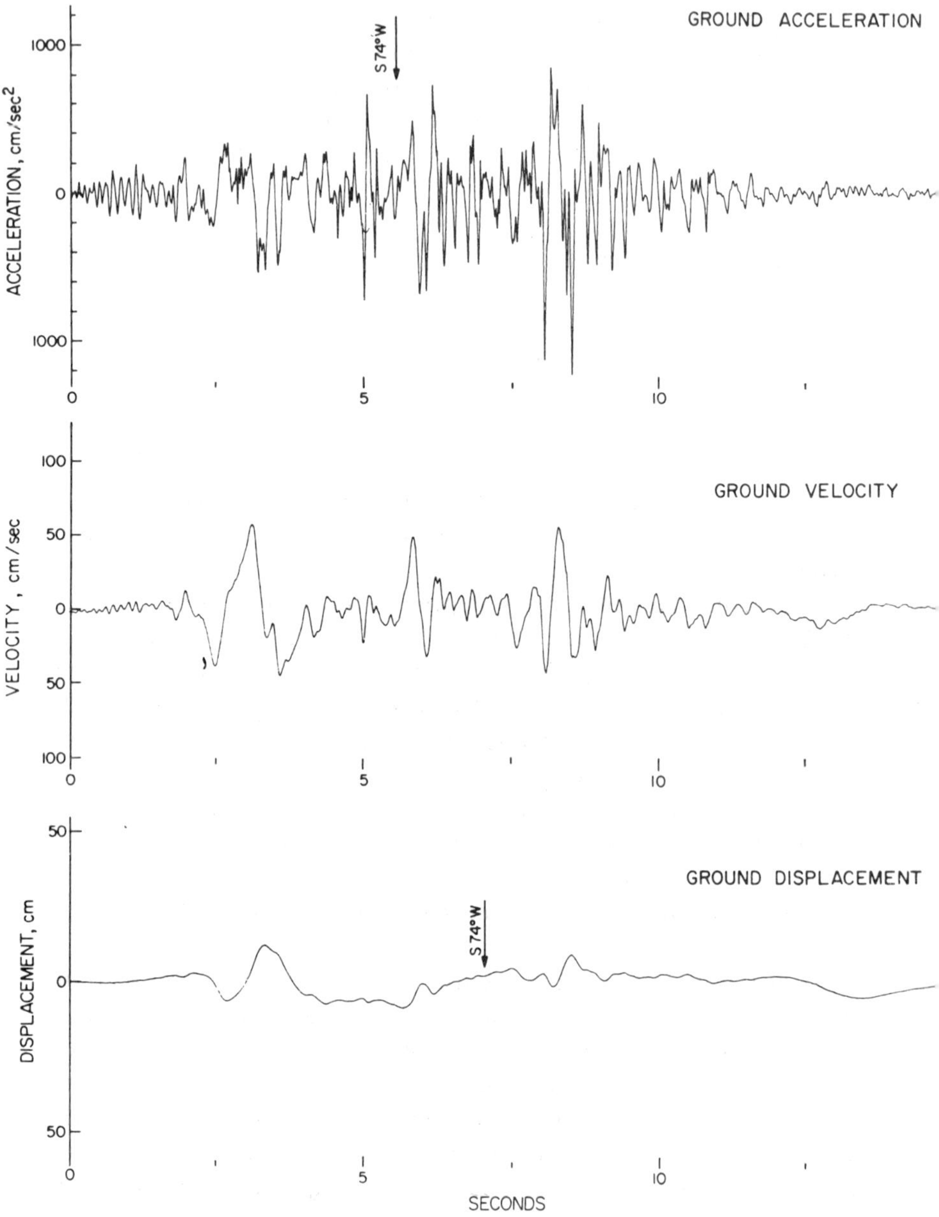

Fig. 4. S74°W motion, Pacoima Dam, San Fernando, Calif., earthquake of 9 Feb. 1971. (After Trifunac and Hudson, 1971; reproduced by permission of the Seismological Society of America.)

strong-motion records available, about 500 from Japan (Committee on Strong Motion Earthquake Observation, 1973) and a similar number from the United States (Brady and Hudson, 1973). An important example of such a record is that taken near Pacoima Dam on sedimentary rock of the San Fernando, California earthquake of 9 February 1971 (magnitude 6.6), as published by Trifunac and Hudson (1971). We reproduce here the record for the S74°W component. The accelerogram is primary; the other dynamic quantities have been deduced from it (Fig. 4).

For engineering applications, the spectrum of the motion is of great importance. Generally, a "response" spectrum is calculated, based upon the response of a single-degree-of-freedom viscously damped oscillator obeying the equation:

$$\ddot{x} + 2\omega\lambda\dot{x} + \omega^2 x = -a(t) \quad (2.33\text{-}1)$$

where x is the relative ground-motion co-ordinate, ω the eigenfrequency, λ the damping constant and $a(t)$ the ground acceleration represented by the accelerogram. Usually, the velocity-response curve $\dot{x}_{max}(\omega)$ is given; it depends only on the damping

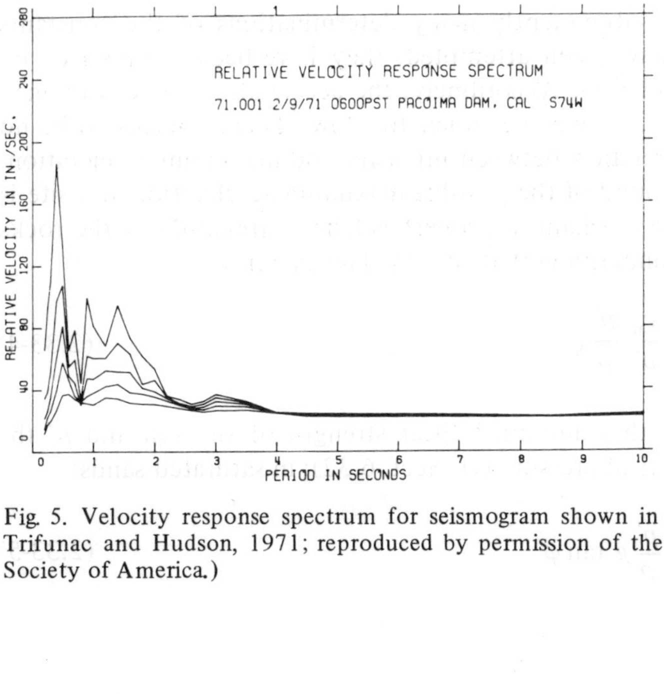

Fig. 5. Velocity response spectrum for seismogram shown in Fig. 4. (After Trifunac and Hudson, 1971; reproduced by permission of the Seismological Society of America.)

parameter λ. Fig. 5 represents the velocity-response spectrum for the accelerogram shown in Fig. 4.

The strong-motion spectra are quite individualistic. One cannot scale upwards from small earthquakes (Tocher, 1973) to large ones; they also shift to lower frequencies as time goes on (Saragoni and Hart, 1973).

Such accurate records as from the Pacoima Dam site are not always available. Attempts have been made to find a relationship between the maximum acceleration a_0 and the maximum intensity I_0 of an earthquake. Generally, these relations are of the form (Båth, 1967):

$$I_0 = c_1 \log a_0 + c_2 \tag{2.33-2}$$

A commonly used relationship of this type is that of Gutenberg and Richter (1956); it is (a_0 in cm sec^{-2}):

$$I_0 = 3 \log a_0 + 1.5 \tag{2.33-3}$$

The values of the maximum acceleration given in Table I have been calculated using this formula.

However, subsequently many determinations of the constants c_1 and c_2 have been attempted; they have been summarized by Ambraseys (1973). Accordingly, the acceleration values obtained from eq. 2.33-3 may be much too low. There appears to be no definite correlation between intensity and maximum acceleration. The upper bound of the possible maximum acceleration at a site is given by the mechanical properties (shear strength!) of the rocks or deposits underlying that site. In clays one has:

$$a_{\text{max}} = \frac{c_\text{u}}{p'} \frac{p'}{p} g \tag{2.33-4}$$

where c_u is the undrained shear strength of the soil and p' the effective vertical pressure (cf. Sect. 6.32); in saturated sands:

$$a_{\text{max}} = \frac{p'}{p} g \tan \varphi' \tag{2.33-5}$$

where φ' is the angle of internal friction in terms of effective stresses, and in dry sands:

$$a_{max} = g \tan \varphi' \qquad (2.33\text{-}6)$$

Therefore, low-plasticity clays are not capable of transmitting accelerations of more than about 0.15 *g*, high-plasticity deposits may transmit to 0.35 *g* and saturated sands up to 0.6 *g*. Higher accelerations may be possible in clean sands and gravel, or, of course, in rock (Pacoima Dam site: 1.25 *g*).

2.34. Effects of local geology

It has been observed that the effects on the ground caused by an earthquake vary considerably with the local geological conditions. Generally, the softer the ground, the more severe are the damage effects. On bedrock, the effects are the smallest. This general pattern was already recognized by Wood (1908) in the 1906 San Francisco earthquake, who found that the damage was most severe on waterlogged land fill, whereas it was relatively minor on bedrock.

The above pattern was confirmed in a study on the damage sustained by the Alaska Railroad in the earthquake near Anchorage of 27 March 1964 (Bonilla, 1970). In areas of exposed till and bedrock, there was hardly any damage, but in areas of young unconsolidated water-laid sediments, much damage occured. The damage also increased dramatically with the thickness of such sediments; below 16 m damage was slight, above 33 m it was severe.

The same pattern of ground motion as a function of local conditions is also observed in effects of more distant earthquakes. A comprehensive study of the effects of local geology on ground motions has been made by Borcherdt (1970) who measured the motions at 37 locations in the San Francisco Bay area that were caused by nuclear test explosions in Nevada, some 500 km away. Again, the same pattern was found for the locations underlain by a layer of young bay sand (or artificial land fill); the maximum intensities increased with thickness of the layer by about 3½ levels on the intensity scale when compared with the intensities on bedrock. The intensities were less on sites underlain by older, more consolidated sediments.

The above observations allow the estimation of relative intensities that are to be expected in an earthquake affecting areas with different local geological conditions. It is usually assumed that the intensity increments over the intensity on granite are 0–1 intensity units on limestones and sandstones, 1 unit on moderately firm ground, 1–2 units on coarse-fragmented, sandy and clayey ground and 2–3 units on loose fill. These values, however, are only guidelines; deviations may be possible (see also below).

Again, it is preferable to deal with quantitative dynamic observables rather than with the somewhat qualitative intensities. The study of Borcherdt (1970) mentioned above gives measurements of maximum horizontal ground velocities. Hudson (1972) and Tanaka et al. (1973) made spectral analyses of strong-motion seismograms at closely spaced sites. Unfortunately, no simple picture emerges. In terms of maximum velocity response (or "amplification factor"), it was found that some stations on rock have greater values than some on alluvium. It may be that the surface and subsurface topography influences the ground motion in a significant manner (Bouchon, 1973). Thus, Kobayashi and Kagami (1973) calculated the influence of multiple reflections and Reimer et al. (1973) showed that local conditions helped to produce the large accelerations (1.25 *g*) observed at the Pacoima Dam site in 1971. Moreover, the impossibility of scaling spectra (cf. Sect.2.33) makes it doubtful that one can determine from the effects of small earthquakes local ground "amplification factors" which would be valid for really large, damaging ones.

2.35. Engineering aspects

It is generally desirable to construct a building in such a fashion that it will not be seriously damaged in an anticipated earthquake. The first problem is to assess which feature of the ground motion may actually *cause* damage. It has been found that ground oscillations act mostly through their velocities at frequencies above the resonance frequency of a building, and through their accelerations at frequencies below the resonance frequency of a building. The

TABLE III

Damage to buildings at high frequencies

Building class	v_{max} (cm/sec)
I. Ruins protected by law; already damaged buildings	0.2
II. Normal houses that are already damaged	0.5
III. Normally built houses (masonry)	1.0
IV. Reinforced concrete structures	1.0–3.0

resonance frequency of a building usually ranges between 2 and 8 Hz (Steinhauser, 1972). In addition, the length of duration of the motions is also important (Bolt, 1973), as well as the nature of the ground motion itself. Thus, it has been found that strong-motion accelerograms in California crossed the zero line 7 to 10 times per second, in Peru the number was mùch higher (Housner, 1973).

High frequencies (8–80 Hz) of ground motion are only seldom found in earthquakes; they are most typical for explosions. Baule (1967) divided buildings into 4 classes and determined the maximum velocities v which will not damage them and which may be used to determine maximum permissible explosions (Table III). The dependence of the maximum ground velocity on the charge L and the distance R is:

$$v_{max} = A\, L^{1/2}\, R^{-n} \qquad (2.35\text{-}1)$$

where $n \sim 1.6$ to 1.7. If v is measured in m sec^{-1}, L in equivalent gelatine (lambrite, etc.) and R in m, then A is around 300, but its scatter is so large that it has to be measured individually in specific cases.

At low frequencies (below 2 Hz), it is, as noted, the acceleration that is primarily responsible for the ground-motion damage to buildings. The greatest-risk frequencies are from 0.3 to 0.6 Hz. However, identical maximum accelerations do not always produce identical damage patterns for buildings. What is important is the coupling between the ground motion and the motion of the particular building in question: the response of the building to the motion of the foundation must be described in terms of the response spectrum of the building to the "input". Here it is often easier to estimate the response of of the building than to predict the proper input (ground motion) that will be experienced during a future earthquake. There is a considerable amount of literature on the calculation of responses and on how to make them as favourable as possible. Problems of this type belong to the field of earthquake engineering, described in special reviews (Housner, 1967) and textbooks (e.g., Wiegel, 1970; Newmark and Rosenblueth, 1971).

2.4. Geographic distribution of earthquakes

2.41. General remarks

Any superficial investigation of earthquakes yields the result that the latter occur more frequently in some areas than in others. In effect, most earthquakes occur in several narrow bands that encircle the globe. The earthquake "proneness" of a region is termed its "seismicity". Naturally, one would like to give a quantitative definition of seismicity, but this is not an entirely simple matter: one has to deal not only with the number of shocks per time unit in a given area, but also with their magnitude or intensity. The problem will be studied more thoroughly in Sect.2.6, where the seismic risk to which a region is subject will be discussed. Here we shall deal only with the geographic (this includes depth h) distribution of earthquake foci. This distribution will also refer to a certain time interval during which these shocks have been observed: even in a discussion about earthquake geography, the time-element cannot be entirely excluded.

For our geographic discussion, we shall first present a broad outline of the global occurrences of earthquakes, then discuss various areas separately, and, finally, try to give some general relations between seismicity and tectonics.

2.42. World seismicity

The world's earthquake epicentres were tabulated in a work of

Gutenberg and Richter (1954) that is still a standard reference. This monograph has been compiled from information about earthquakes that occurred since the introduction of a reasonably accurate instrumentation, i.e., about since 1904, to 1952. Richter (1970) has cautioned that no attempt was made by the authors to include epicentres exhaustively below magnitude 7.0; the attention was directed more to indicate minor active zones outside the main belts. Nevertheless, the picture of seismicity obtained by Gutenberg and Richter (1954) is in essence still valid today. The work of Gutenberg and Richter was continued by Rothé (1969a) until 1965; Barazangi and Dorman (1969) also compiled global seismicity maps for 1961–1967 from ESSA, U.S. Coast and Geodetic Survey epicentre data. The general picture already obtained by Gutenberg and Richter in 1954 was thereby confirmed. A picture of world seismicity is shown here in Fig.6.

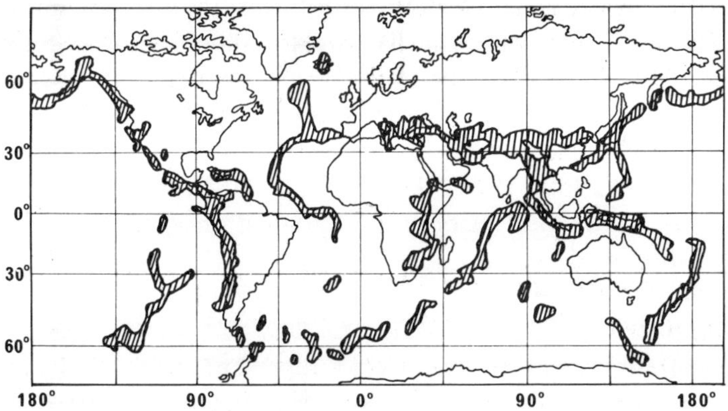

Fig. 6. World seismicity; areas where earthquakes occur with magnitudes > 6 are shown shaded.

Accordingly, most epicentres occur along oceanic ridges (and corresponding continental rifts), along continental transcurrent faults and on active (with island arcs) continental margins.

The depth h of foci reaches to about 600 km, but other than crustal earthquakes ($h < 30$ km) occur only, with few exceptions, in narrow planar zones dipping beneath island arcs.

2.43. Types of regions

A. Introductory remarks

We shall now discuss the seismicity of single types of regions, and we will try to point out their characteristic features. The information is gleaned from the world seismicity surveys mentioned in Sect. 2.42 as well as the individual studies mentioned below. Only a broad outline will be presented; for details, the reader is referred to the cited monographs.

B. Continents

The central shield areas of continents (the Canadian shield, the Brazilian shield, the Fennoscandian shield, the Angara shield, the African mass, western Australia and Antarctica) are the least seismic regions of the world; of these, the Antarctic (Evison, 1967b) is particularly free of earthquakes. In the Fennoscandian shield, which is rising because of isostatic uplift, earthquakes are more frequent (Båth, 1956); the same is true for the Canadian Shield (Smith and Milne, 1969). For continents in general, the earthquakes occur at localities of recent tectonic disturbances. Careful studies have been made about disturbances in Europe and in the U.S.S.R. (Karnik 1960, 1971; Akad. Nauk SSSR, 1961). Mountains, rifts and continental margins are the most important areas of disturbance discussed.

C. Mountains

The mountain ranges of the world represent areas of major tectonic disturbances. They form two large belts, one around the Pacific and the other through the Alpine and Himalayan systems to Oceania. Seismically, the Pacific belt, which will be discussed in connection with continental margins, is somewhat different from the epicontinental Alpine–Himalayan system. It is associated with many shallow and intermediate-depth earthquakes, some of large magnitude. However, only very few deep-focus earthquakes have been observed in this belt. Minor seismic activity is associated also with older mountain ranges, such as the Ozarks and the Appalachians.

D. Rift zones

The counterpart of mountains are rift zones. There is a large rift-zone system dissecting the eastern parts of Africa whose seismicity has been studied extensively (De Bremaecker, 1959; Wohlenberg, 1968; Molnar and Aggarwal, 1971). Minor rift zones are found in the Lake Baikal region and in the Rhine graben (Rothé and Schneider, 1968). The earthquakes occur in the rift zones; they are, however, less severe than in the mountain areas. Nevertheless, magnitudes up to about 8 have been observed in the Rhine graben (Basel, 18 Oct., 1356, $I_0 = 10$) and in the African rift valleys.

E. Continental margins

Continental margins represent some of the most seismically active regions of the world. This applies particularly to the Circum-Pacific belt which is characterised by volcanism, mountains and island arcs. This belt accounts for a large majority of the shallow and intermediate global earthquakes, and for almost all deep-focus earthquakes. These earthquakes are often concentrated on visible faults (cf. the San Andreas fault in California).

F. Sea-floor

As with continental shields, the true abyssal plains in the global oceans appear to be seismically inactive zones: as on continents, seismicity is confined to narrow, disturbed bands, notably the mid-ocean ridges and the archipelagos (Sykes, 1963; Stover, 1966).

G. Mid-ocean Ridges

Like the mountain chains, mid-ocean ridges also form a worldwide system: it starts with the Mid-Atlantic ridge, which continues from the Arctic Ocean to Tristan da Cunha, then it swings around Africa and connects with the Madagascar and St. Paul ridges in the Indian Ocean. Finally, it continues to the south of Australia and connects with the principal ridges in the Pacific Ocean.

The structures of the ridges is quite characteristic: lateral offsets are effected by transform (not transcurrent) faults; a rift follows the ridges at the crest so that the latter appear as structures caused by upwelling of material from the mantle of the earth. The rifts are similar to, and probably genetically connected with, continental rift valleys. Correspondingly, their seismicity is also very similar to that on continental rifts: occurrence of many shallow, relatively minor earthquakes. Earthquakes are particularly frequent on the transform fault lines.

H. Archipelagos

The seismicity of archipelagos is similar to either that of continental margins (island arcs) or to that of a mid-ocean ridge, depending on whether the islands are peaks of a mid-ocean ridge or extensions of a precontinental island arc (like the Bonin–Marianas archipelago).

2.5. Temporal distribution of earthquakes

2.51. Introduction

The geographic distribution of earthquakes discussed so far represents only a very poor characterization of the seismicity of an area. It is not only important to know that earthquakes occur in an area, but also how frequent they are.

It appears that there is a rather simple relation between the frequency and the magnitude of the earthquakes which occur in an area. Earthquakes, however, have a tendency to cluster in groups (mostly as aftershock sequences) so that characterization of their occurrence as a time series is not very easy to achieve.

Finally, the seismic strain or energy release for an area can be calculated and maps of "tectonic flux" can be constructed.

These questions will be analyzed below in detail.

2.52. Magnitude–frequency relations

It was recognized fairly early that the number N of earthquakes occurring during a given time interval in a given area decreases with their magnitude M. Gutenberg and Richter (1949) postulated the relation:

$$\log N(M) = a - bM \tag{2.52-1}$$

If this relation can indeed be established for an area, then the likelihood of a strong earthquake can be estimated from observations of weak earthquakes over a limited period. The formula is valid only for $M \leqslant M_{max}$ for a given area.

It is clear that the quantity b in eq. 2.52-1 is independent of the time interval or of the size of the area under consideration. There is some controversy whether b is a universal constant or varies strongly from area to area. Miyamura (1962) collected the available evidence and found that very low b values (0.4–0.6) were reported for old shields, low values (b = 0.6–0.7) for continental rift zones, medium values (b = 0.7–1.0) for the Circum-Pacific and Alpine orogenic belts and high values (b = 1.0–1.8) for oceanic regions. It should be mentioned, however, that Riznichenko (1969) has strongly questioned Miyamura's results and claims that b is in fact a universal geophysical constant that is determined by the fracture mechanics of the material in question and which should be the same everywhere.

The statistical relation eq. 2.52-1 can also be represented in other forms. Combining it with eq. 2.24-2 and eq. 2.24-3 one can write (Riznichenko, 1965a):

$$N = A \cdot 10^{-\gamma(K-K_0)} \tag{2.52-2}$$

where A and γ are now the parameters replacing a and b in eq. 2.52-1; K_0 is an additionally assigned number. The coefficient γ is independent (like b above) of time interval and size of area; in effect, it is simply:

$$\gamma = \frac{b}{\beta} \tag{2.52-3}$$

where β is the coefficient defined in eq. 2.24-3, i.e., by the magnitude–energy relation. The presentation of the earthquake frequency law in terms of the focal energy (E or K) is limited because of the difficulties of determining the latter. Whilst this is *physically* a most sensible procedure, an uncertainty is introduced in practice because of the poor definition of the coefficient β. Nevertheless, similar studies of a seismic regime can be made by using the coefficient γ (eq. 2.52-2) instead of b (eq. 2.52-1).

We thus have a means to characterize "seismicity" quantitatively. Eq. 2.52-1 can be written:

$$\log N = a - b(M - M_0) \tag{2.52-4}$$

where M_0 is some conventionally fixed magnitude. The above relation is completely determined if the constants a and b are known. As noted, to make comparisons, the size of the area and the time interval must be standardized. Then a represents the seismic activity, and b a (possibly universal) tectonic constant. There may, in effect, be an additional parameter necessary for determining the seismicity of a region: the magnitude M_{max} of the maximum possible earthquake.

In terms of energy rather than magnitude (using eq. 2.52-2) the formula equivalent to eq. 2.52-4 is:

$$\log N = \log A - \gamma(K - K_0) \tag{2.52-5}$$

where again K_0 is some conventionally fixed energy-class value; A is then the activity and γ the "tectonic" parameter.

The various coefficients characterising seismicity may vary not only from area to area, but also with depth. Butkovskaya and Kuznetsova (1971) found a decrease in the slope of the recurrence rate curves with focal depth in Uzbekistan.

Similarly, there have been claims that *secular* variations of b (or γ) have been discovered (Ikegami, 1968; Lukk and Ponomarev, 1972).

In addition to the above parameters, the magnitude M_{max} (or energy parameter K_{max}) of the *maximum* earthquake that could occur in an area must be specified. Riznichenko (1964) obtained a correlation for K_{max}:

$$\log A = \log \alpha + \beta(K_{max} - K_\alpha) \tag{2.52-6}$$

where the coefficients (assuming $K_0 = 10$, time unit = 1 year, area unit = 1,000 km^2) $\log \alpha = 2.84$, $\beta = 0.21$, $K_\alpha = 15$ were determined by a least squares procedure from observations for Central Asia. For the Crimea Riznichenko et al. (1969) used $\log \alpha = 2.63$; the other parameters had the same values as mentioned above.

The last formula enables one to draw maximum possible earthquake maps for an area. However, the uncertainty regarding coefficients is large.

2.53. Aftershock sequences

It is a well-known fact that large earthquakes seldom occur alone: they are associated with a series of "foreshocks" and "aftershocks". A comprehensive statistical study on the phenomenology of aftershock sequences has been made by Ranalli (1969) and by Ranalli and Scheidegger (1969). Accordingly, the statistical laws postulated earlier by various authors have been fully confirmed by a detailed analysis of 15 aftershock sequences. These statistical laws are the following:

(1) Aftershocks are generally crustal events (Page, 1968); deep aftershocks occur only seldom: the deepest sequence (Dzhurm region, March 14, 1965; Lukk, 1968) has a depth of about 240 km.

(2) The frequency N (per unit time) of aftershocks above a certain cut-off magnitude within the same sequence decays with time according to the law first postulated by Omori (1894):

$$N(t) = \alpha t^{-\beta} \qquad 1 \text{ day} \leqslant t \leqslant 100 \text{ days} \qquad (2.53\text{-}1)$$

where β is approximately equal to, or slightly greater, than unity. The values found by Ranalli in the 15 sequences analysed by him range from 0.5 to 1.9. The values for α depend on the cut-off magnitudes for the observations used. Omori's law is a statistical law; i.e., the time unit chosen for N should be sufficiently large; fluctuations around the expected value do occur.

(3) The average magnitudes of aftershocks (above a certain cut-off magnitude) show stability in time to the end of the sequence (Lomnitz, 1966a). Again, this is a statistical law implying,

e.g., that the mean magnitudes of groups of, say, 10 successive aftershocks, will show only random fluctuations above the overall mean magnitude of the entire sequence.

(4) The magnitude–frequency distribution in an aftershock sequence is of the same type as that postulated by Gutenberg and Richter (1949) for earthquakes generally, viz.:

$$\log N = a - bM \tag{2.53-2}$$

The parameter b is found to change for different aftershock sequences; in the 15 sequences studied by Ranalli (1969), it was found to range from 0.44 to 1.36. Whether these differences are significant is, as for earthquakes in general, still an unsolved question.

2.54. Earthquakes as a time series

The number of earthquakes (above a certain cut-off magnitude) per time unit in a certain area (this may be the whole world) forms a "time series". It is interesting to study the statistical properties of this time series. In practice, the time unit is the day, but motions occurring on the average over 25 or 50 days are usually considered rather than single-day observations. From the U.S.C.G.S. earthquake catalogue the nature of the time series obtained as described above for the whole world from 1961 to 1969 is shown in Fig. 7. Similar time series can be obtained for individual areas.

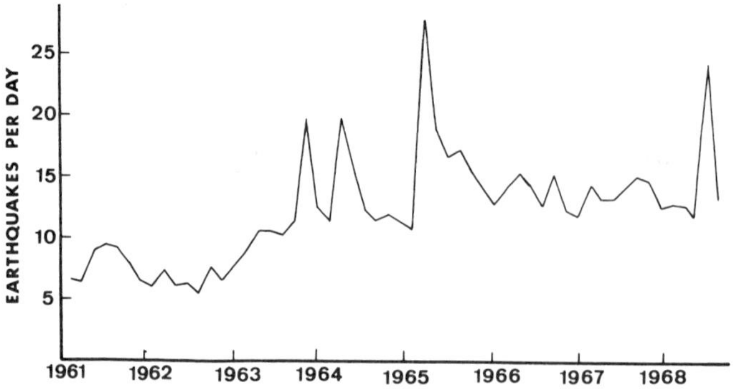

Fig. 7. Average number of earthquakes per day in the U.S.C.G.C. catalogue using 50-day averages. (After Shlien and Toksöz, 1970.)

Because of the occurrence of aftershocks, it is clear that the time series mentioned above must show clusters. The peaks seen in Fig.7 indeed represent the aftershocks of great earthquakes. With increasing cut-off magnitude, the peaks tend to disappear. Ben-Menahem (1960) concluded already some time ago that, if the obvious aftershocks are removed, the time series is then that corresponding to a Poisson process. For the latter, the probability $p(j;t)$ of j earthquakes occurring in the time interval t is given by:

$$p(j, t) = \frac{(kt)^j}{j!} e^{-kt} \tag{2.54-1}$$

The recurrence time, then, is exponentially distributed and the mean and variance of the number of events per unit time are both equal to k.

Because of the obvious grouping of earthquakes, most tests of a time series of earthquake occurrences of low magnitudes yielded results which did not fit the Poisson process very well (Vere-Jones et al., 1964; Knopoff, 1964; Isacks et al., 1967; Ferraes, 1967). Thus, various clustering models have been proposed (Vere-Jones, 1966), of which a most elaborate one is that suggested by Shlien and Toksöz (1970) embodying a "generalized" or compound Poisson process: the clusters are Poisson-distributed in time, and the events in each individual cluster are distributed according to a negative power-law distribution. In this fashion, a fair fit between the model and the observations can be obtained. In effect, this procedure also amounts to a "removal" of the "associated" earthquakes from the "main" shocks, or clusters. The clusters are the primary events; they are, then, Poisson-distributed.

Attempts have also been made to detect periodicities in the occurrence of earthquakes which could be related to tidal or cosmic influences. Only Tamrazyan (1968) believes that there is a correlation with lunar and solar forces, none of the other investigators (Knopoff, 1964; Gayskiy, 1967) were able to detect any periodicity. Gayskiy (1967), after removing the aftershocks, found some discrepancy in the resulting time series with that corresponding to a Poisson process. He interpreted this as a "tendency" toward periodicity.

In view of the above, and with the information available to date, it can probably be stated that the earthquake occurrences correspond fairly closely to a time series engendered by a Poisson process, on the condition that evident "clusters" are removed; i.e., clusters are counted as single "events".

2.55. Strain and energy release

As was noted in Sect.2.24, the square root of the energy radiated in an earthquake is proportional to the strain released in that earthquake. This idea was used for the first time by Benioff (1951) to plot curves of cumulative strain release for a given period for a given area (which may be the whole world).

The possibility of plotting cumulative strain-release curves has been used in connection with the characterization of single areas in the study of aftershock sequences, etc. In conjunction with source models, inferences regarding the rheology of the tectonically significant layers of the earth can be drawn (Ranalli and Scheidegger, 1969).

As noted, the amount of strain release in a given area can be used to characterize the seismicity of that area. Naturally, the data must then be referred to a standard area size and to a constant time period. In order to avoid the introduction of elastic constants, the cumulative strain release may be stated as the number of equivalent earthquakes of a pre-set magnitude M_x. This method was employed by Allen et al. (1965) to characterize the seismicity in southern California. Their map shows interesting correlations with the tectonic fault lines of the area. Milne (1967) used the same•method to characterize the seismicity in Canada.

A notion related to strain release is the tectonic flux h of an area. This is simply the secular rate of seismic strain release per unit area A expressed in erg½/time. An arbitrary constant a is used to achieve a reasonable scale (St. Amand, 1956):

$$h = \frac{a}{A}\,\frac{1}{(t_1, t_2)} \int_A \int_{(t_1,t_2)} E^{1/2}\,\mathrm{d}A\,\mathrm{d}t \qquad (2.55\text{-}1)$$

where (t_1, t_2) is the time interval concerned. The seismic energy, of course, is calculated according to one of the magnitude–energy relationships. In this fashion, Hédervári (1971) obtained a map of the tectonic flux in Japan.

Instead of the strain release, the energy release has also been plotted, particularly by Båth (1953) in Fennoscandia and by Tsuboi (1965) in Japan. Again, a characterization of the seismicity of a region is obtained in this fashion.

2.6. Seismic risk

2.61. Introduction

The seismic risk of an area is represented by the danger of earthquake damage that structures are subject to. The risk is then connected with the potential ground motion; it is a concept related to the potential earthquake intensity or to the potential accelerograms that may be recorded in a given locality with a given frequency.

It should always be kept in mind that the very concept of "risk" is statistical in its nature. A design based on such a risk can, therefore, never be absolutely "safe". Thus, the problem is one of optimization which may imply using such procedures as putting a monetary value on the human lives that may be lost in a potential disaster and weighing this cost against the costs of various designs. Questions of ethics naturally enter into such considerations (e.g. Rosenblueth, 1973).

In the previous sections, the means have been developed to discuss the chain of physical arguments that leads to seismic risk and its characterization in greater detail. To conclude the chapter we shall discuss some problems concerning the possible prediction of earthquakes.

2.62. Characterization of the seismic risk

The seismic risk can be characterised by the number B of shocks that may be expected to occur at a certain locality in a given time unit with an intensity I or greater:

$$B = B(I) \tag{2.62-1}$$

This quantity has been termed the "shakeability function" of an area (Riznichenko, 1965b). The inverse (referring to the chosen time unit) is the recurrence time T for shocks of a given intensity I or greater:

$$T = T(I) = 1/B(I) \tag{2.62-2}$$

The primary *physical* cause of the ground motion is not necessarily located at the geographical point at which the seismic risk is to be estimated: the primary causes for the creation of the seismic risk are the potential earthquake foci that may arise within a certain vicinity of that point. The first step, therefore, is to estimate the likelihood of the occurrence of an earthquake focus within the general vicinity of the point under consideration as a function of its magnitude; then, the effect of such possible foci, as far as the intensity at that point is concerned, has to be summed up to evaluate the seismic risk at that point.

Since the maximum intensity I_0 is a linear function of the earthquake's magnitude, one would expect that the general form of the frequency relation (eq. 2.52-4) is retained in the summation; hence, as a first approximation, one can try to set:

$$\log B(I) = a - b(I - I_c) \tag{2.62-3}$$

where I_c is some conventionally fixed intensity value.

Therefore, one again has to determine only two parameters, a and b (these are different from the corresponding parameters in the magnitude–frequency law) to characterize the seismic risk at a point. In addition, the maximum intensity I_{max} that is possible at a point may also have to be specified. Since the intensity is affected by local ground conditions, a reduction to some standard ground has to be made as well.

The practical calculation of the seismic risk parameters at a certain point is thus a procedure that involves several steps. First, the seismicity for the whole surrounding region has to be determined; i.e., the magnitude–frequency relations for the *foci* that can be expected in the vicinity of that point must be established. Then the effect upon the expected intensity at that point must be calculated by a summation (integration), by the use of a formula similar to eq. 2.32-3 for the intensity–distance relation. For some simple cases (distribution of foci homogeneous on a plane, etc.) Riznichenko (1965b) has given analytical expressions; in real cases, however, the integration has to be performed numerically. The results of such calculations are commonly represented in a series of charts in which the shakeability is given as $T(I_n)$ where T is the estimated recurrence time in years of ground shakings with an intensity greater than or equal to I_n. The chart of maximum possible earthquakes has to be determined from eq. 2.52-6.

The characterization of a seismic risk given above is based on the concept of the earthquake intensity. As pointed out already on several occasions, the concept of intensity is a somewhat qualitative one. For the engineer, a quantitative estimate of the accelerogram of an earthquake that may be expected at a given locality would be more desirable, inasmuch as he could then use this as a basis for his designs. Thus, one begins again with the seismicity of a region characterized by the parameters a and b in eq. 2.52-1. For the potential earthquakes of an area, typical accelerograms are then selected from the catalogues mentioned in Sect.2.33 allowing, as best as one can, for such factors as the type of ground in question and the distance from the expected focus.

The above procedures are undoubtedly physically satisfactory, but difficult to carry out. Hence, much simpler procedures than those sketched above have also been used to characterize the "seismic risk" at a point; usually by concentrating on focal rather than on ground properties. The strain release, the energy release and the likelihood of earthquakes exceeding a certain magnitude (rather than intensity) have all been used for this purpose. Ranalli (1972) applied statistical methods to the analysis of interoccurrence times x_M between earthquakes above a certain magnitude M in analogy with the procedure commonly applied to the analysis of volcanic eruptions (cf. Sect.3.52). If these interoccurrence times x_M are assumed to be Poisson-distributed, then the parameter k of the Poisson distribution (cf. eq. 1.32-1 with $j = 1$) can be used for the characterization of the seismic risk of an area. The magnitude–maximum intensity relation gives then a means for estimating the damage to be expected and the probable frequency of its recurrence, without having to take recourse to the elaborate means sketched earlier in this section. However, in this fashion only focal quantities are obtained; for the real estimation of the risk at a point, a summation (integration) over the possible sources in its vicinity will always have to be performed.

2.63. Seismic zoning

For the establishment of building codes, it is evident that it is necessary to create a map of a region that will indicate the degree of seismic danger.

We have seen above that at least 3 parameters are required to determine the seismicity (referring to epicentres) of a region: the constants a and b in the recurrence graph, and the magnitude of the maximum possible earthquake M_{max}. As noted it may be hoped that similar laws also refer to the intensity I, so again 3 parameters would be needed to determine the "shakeability" of a region. In essence, the procedure to do this has been described in the last section; the maps of seismic risk obtained according to the procedures of Sect. 2.62 precisely represent zoning maps. However, in this fashion, only an overall zoning is obtained; locally, the geological conditions (water-logged sediments or bed-rock) can effect the variation of intensity values of up to 3½ scale units, and local topography can also have an influence. Equally, the presence of an active fault (Ziony et al., 1973) can greatly increase seismic risk. In the practical procedure, intensity values (also in the raw data) have to be reduced and calculated for some "standard" ground; the deviations from these values, then, have to be estimated after allowing for the geological and topographic peculiarities of a particular location.

Even with the above adjustments, zoning for seismic risk is an uncertain procedure (Housner and Jennings, 1973). In principle, all zoning maps represent some more or less sophisticated extrapolations from past history. Whilst the general level of the seismic activity in a region can thereby be ascertained, the particular locations of a future large earthquake cannot be predicted (see also the next section). In a seismically active area, large earthquakes do not necessarily occur in the region where most small earthquakes do; and they may occur in a location where no large earthquakes have been observed for long periods. For example, on October 5, 1948, a large earthquake (M = 7.6) struck near Ashkhabad, in a region where no strong earthquakes had been experienced for several centuries; although it was known, from the general proximity of Iranian earthquakes, that the area might be seismically active. Nevertheless, the exact location and the magnitude of the Ashkhabad earthquake came as a surprise.

Microzoning, to date, is of doubtful value (Hudson, 1972; Housner and Jennings, 1973; Hudson and Uwadia, 1973). There are so many factors that affect the severity of the local ground motion that the most that can be done is to state whether a site is bad, average or good. One cannot make a microregionalization of the bad zones, but can only indicate where it might be very bad (catastrophic) due to potential sliding, slumping or liquefaction. Correlations with data from small earthquakes are meaningless because spectra cannot be scaled (see Sect.2.33).

2.64. Earthquake prediction

It would undoubtedly be very desirable if the time of occurrence, the locality and the strength (preferably the intensity at the surface) of an earthquake could be predicted: if an earthquake cannot be prevented, at least people could be warned to leave their houses and move to the safety of open spaces in order to minimize the death toll. Unfortunately, to-date it is not possible to do this in spite of much serious effort that has been made to achieve this in various countries, particularly in Japan (Alsop and Oliver, 1969), the U.S.S.R. (Akademiya Nauk, 1970) and the U.S.A. (ESSA, 1970); however, success may not be very far off.

We shall briefly review the various possibilities that have been invoked.

A. Statistical extrapolation

The simplest procedure for predicting earthquakes would undoubtedly be an extrapolation of the series describing their occurrence (e.g., time-intervals between large earthquakes in an area) into the future. This approach has been tried particularly by Lomnitz (1966b, 1967). However, it appeared that this approach is basically useful only for risk estimates.

B. General premonitory indications

The search, therefore, has been for indications that may herald an earthquake. Forwarnings may be gleaned from tilts and strains in the epicentral region, from a change in micro-earthquake activity, and from changes in the physical properties of rocks such as resistivity and magnetic properties leading to the "seismomagnetic" effect (Båth, 1966b; Press and Brace, 1966; Rikitake, 1968) and to changes in telluric currents. A generalized "forerunner curve" has been compiled by Myachkin and Zubkov (1973) which indicates that the earlier forerunners will occur the larger the earthquake is. The corresponding equation is:

$$\log \Delta T = 0.48\ K - 7 \qquad (2.64\text{-}1)$$

where ΔT is the appearance time of forerunners in years before the main earthquake and K its energy-class (cf. eq. 2.24-2).

C. Strain build-up

Anomalous deformations prior (hours or days) to large earthquakes have been reported on several occasions; they may be detected as changes of sea level, or horizontal and vertical shifts revealed by accurate surveys. Such shifts may be caused by an acceleration of creep phenomena (Tomashevskaya and Kamidullin, 1972) leading to the earthquake. The unequivocal premonitory significance of such changes, however, is by no means certain yet (Press and Brace, 1966; Aki, 1972b). The strain build-up may express itself also in a change of mechanical rock properties, manifested in a change of wave propagation velocity, particularly evident in the ratio of pressure to shear wave velocity (Nur, 1972).

Nur (1973) observed that the ratio v_p/v_s of the velocities of P to S waves has a characteristic course before the onset of a major earthquake. The wave velocities are measured during the occurrence of minor earthquakes. Then the time-evolution of the ratio v_p/v_s is found to have a characteristic bay-shape: it decreases at first, then mounts rapidly back to its original value, rises substantially above it until the earthquake occurs, and drops finally back to its "normal" value. Depending on the magnitude of the major earthquake, the "bay" may last some days, months or years. The pattern of "bays" appears to occur consistently as a precursory event before thrust-type earthquakes with a focal depth of not more than 10 km. It can be hoped, therefore, that these types of earthquakes can be predicted if the characteristic "bay" is observed. This prediction technique has been applied successfully by Aggarwal et al. (1973) to the prediction of shallow earthquakes occurring in New York State.

Regarding a physical explanation of the bay-phenomenon, Nur (1973) noted that generally the wave velocity is affected by porosity and saturation. The crustal rock, even granite, is slightly porous; in the "normal" state it is fully saturated with pore fluid (water). Just before fracture, the porosity increases because of the occurrence of stress dilatancy (cf. Sect.4.15). During the occurrence of dilatancy the rock becomes undersaturated since, because of the low permeability, it will take some time until the full saturation state is again reached. When this occurs, the rock may finally fracture and lead to an earthquake. In this fashion, the bay-phenomenon can be explained.

D. Small-earthquake activity

It has also been noted that changes in micro-earthquake activity precedes some earthquakes, usually expressing themselves by the arising of "foreshocks" (Mogi, 1962; Alexander, 1968). However, Oliver (1966) noted that micro-earthquake activity is primarily related to large earthquakes that occurred 10 years ago rather than to future earthquakes.

E. Resistivity

Electrical properties of water-saturated rocks change if the rocks are strained; thus, a change of resistivity may be expected when the rock is being stressed to the breaking point. However, so far field tests (Rikitake and Yamazaki, 1967) have still been inconclusive.

F. Seismomagnetic and telluric effects

The reason for suspecting a seismomagnetic effect lies in the observation that the magnetic susceptibility changes if material is stressed. Thus, a change in the magnetic field may accompany a dangerous stress build-up. Rikitake et al. (1967) reported on a series of proton precession magnetometer observations in Japan during the occurrence of an earthquake swarm. Again, the results were inconclusive.

Somewhat better correlations were obtained between strong earthquakes and telluric currents in a study of conditions in Kamchatka (Sobolev et al., 1972), but such currents do not seem to be sufficient for earthquake predictions either.

G. Artificial release of earthquakes

Related to the problem of prediction of an earthquake is that of artifical release. The problem has been reviewed by Båth (1970). Accordingly, artificial loads causing stresses at or near the surface of the earth can engender earthquakes. Large reservoirs filling with water due to a dam's construction have been known to activate faults in their vicinity and cause earthquakes. (Cremasta dam in Greece, $M = 6.2$ in 1966; Kariba dam in Zambia/Rhodesia, $M = 6.1$ in 1963; and probably Koyna dam in India, $M = 6.5$ in 1967.) After the complete impounding of the water, the seismicity continued for some time and then gradually decreased, evidently until a new equilibrium was reached. Similarly, the injection of wastes into wells (3,670 m deep) near Denver, Colorado, represents a famous case of the artificial creation of (frequent small) earthquakes (up to 120 per month; maximum magnitude $M = 4.3$). It has been suggested that by fluid injection the stresses, which are building up, could be released before they are so large that a disastrous earthquake results. On the other hand, large nuclear explosions, while they may trigger small "aftershocks" (the biggest about one magnitude-unit lower than the main explosion), do not seem to have much effect on the seismicity of an area. It is, at most, conceivable that a large earthquake, which would be occurring soon, could be triggered by an explosion.

Chapter 3. VOLCANIC ERUPTIONS

3.1. Phenomenology of volcanic eruptions

3.11. Definition

The term "volcanic activity" encompasses all phenomena that are connected with the ascent of hot materials from the interior of the earth to the surface: these materials include gases, hot water (steam) and lava (molten rock).

When hot materials ascend to the surface of the earth the confining pressure drops, so that gases that were previously in solution escape from molten rocks. The sudden appearance of molten rock from a single centre (vent) or a fissure on the earth's surface, combined with the sudden release of highly compressed gases to atmospheric pressure, produces the appearance of an "eruption". Generally, a cone of debris material has been built up earlier by previous eruptions (giving to volcanoes a characteristic appearance), the lava flows down the slopes, rocks are thrown out of the vent as "bombs" and the escaping gases may carry small to very fine particles ("ashes") over large distances. The ashes may be glowing hot and roll in a burning cloud (nuée ardente) along the surface. Sounds accompany the eruption (Richards, 1963; Richards et al., 1962). Volcanoes may be dormant for hundreds of years and then suddenly erupt again or they may form in an area where there have never been any volcanoes previously (Paricutin, Surtsey) although they seem to be confined to certain general regions of the world. An eruption may last from a few hours to a few months; dormant periods may then again follow. The pertinent features of volcanic eruptions have been described on many occasions in the literature. A notable collection of the physiographic features associated with volcanic eruptions has been made in the form of a photographic atlas by Green and Short (1971). Of the many textbooks on volcanology, we only mention the monograph by Rittmann (1960). The accounts of individual eruptions are, of course, legion. We shall refer to them when needed.

3.12. Classification of volcanic eruptions

The characteristics of a volcanic eruption are very much determined by the amount and pressure of gas that is released from the molten rock and by the viscosity of the lava. In the viscosity of a lava, one distinguishes pahoehoe (or ropy) lava as the most fluid, aa lava as the intermediate and block lava as the most viscous. Pahoehoe lava may change into aa lava farther from the source (the loss of volatile materials causing the increase in viscosity); block lavas are generally more silicic in composition than the other types. It is difficult to give actual values for the viscosities encountered, as they are highly temperature-dependent. The classification given above is therefore only a relative one. Nevertheless, it should be mentioned that Rittmann (1960, p. 186) conjectured from observation that very fluid lava with a high gas content may have a viscosity close to that of water (0.02 Poise) at high temperature. For melts of basalt, orders of magnitude of 10^2–10^4 Poise were found in the laboratory, but such melts do not contain any volatile substances (see also Sect.3.42). When a lava is cooled, a glassy or polycrystalline structure may be the result, depending on the initial composition.

As with the viscosity, the gas quantity and pressure in an ascending magma may also be low, medium or high. The types of eruptions, named after characteristic volcanoes and regions, which have been described in the literature, can then be classified as shown in Table IV (modified after Hédervári, 1972, who followed in part earlier authors). The Hawaiian and Icelandic type eruptions show continuous activity; the Icelandic type generally originates in extended fissures, while the Hawaiian type originates in fissures or in centres. The Strombolian type is more violent than the previous ones and produces clouds, the Vesuvian type even more so. The Vulcanian type is characterized by vertical explosion clouds; in an eruption of the Perretian type these clouds are gigantic and are connected with ash falls (Vesuvius 79 a.d.). The Merapi type is

TABLE IV

Classification of volcanic eruptions
(Modified after Hedervari, 1972)

Lava	Gas pressure		
	Low	Medium	High
Fluid	Icelandic Hawaiian	Strombolian	Vesuvian
Intermediate		Vulcanian	Perretian
Viscous	Merapian	Vincentian	Peléean

characterized by glowing clouds which become glowing avalanches (nuées ardentes) in the eruptions of Vincentian type and giant glowing avalanches in the eruptions of Peléean type. The most fearful effects for mankind are produced by a combination of high gas pressure and high viscosity of the lava.

3.2. Volcanicity of the earth

3.21. The geographic distribution of volcanoes

Volcanoes, like earthquakes and mountain chains, do not occur

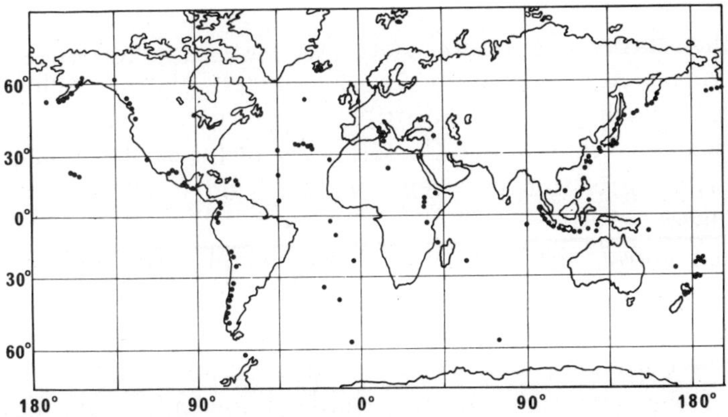

Fig. 8. Distribution of volcanoes.

at random on the earth. In fact, they are confined to definite "volcanic regions".

Characteristics of individual volcanoes and their locations have been described on many occasions in the literature. This information has been collected by the International Association of Volcanology (1953/67) in a catalogue of 20 volumes about the active volcanoes of the world.

The location of the most important volcanoes is shown in Fig.8. Naturally, not every known volcano could be plotted in this figure; nevertheless, the general character of the geographic distribution of the volcanoes is immediately clear.

The present distribution of volcanoes may not always have been the same. A study of palaeovolcanology has been made by Cook (1966). However, for the discussion of potential present-day catastrophes, this is of little significance.

3.22. Correlation of volcanic activity with tectonics.

An inspection of Fig.8 shows that the volcanoes predominate in regions which generally exhibit present-day tectonic activity as well. There is thus a correlation between volcanicity and tectonism. We note, therefore, that many volcanoes occur in the Circum-Pacific belt of orogeny, the mid-oceanic (Iceland, Azores) and epicontinental rift regions (Africa) and, to a lesser extent, in the Alpine–Himalayan mountain region (including the Italian volcanoes).

It is, therefore, only natural that there is also a correlation between volcanicity and seismicity of a region; this is evident when Fig.8, which shows the distribution of volcanoes, is compared with Fig.6, which shows the seismicity of the earth. Similarly, there is a correlation between volcanism and terrestrial heat flow (Boldizsár, 1970).

As noted above, the correlation between volcanism and tectonism is only a *general* one. The details of the structure of a tectonically active region are often very complicated. A thorough study of the correlation between the distribution of foci, the seismicity and the volcanicity of the Kurile–Kamchatka zone has been made by Tokarev (1970); Sugimura (1963) made a similar study of the east and west Japan volcanic belts. The picture that evolves from observations in these regions is that, starting from the oceanic side and proceeding to the continent, one first finds a deep ocean trench, then a line of volcanoes from which a zone of earthquake foci starts that dips downward toward the continent. This type of structure is presumably associated with a subduction zone of an oceanic plate spreading toward and disappearing under the continent as postulated in the new global tectonic theory (see e.g. Johnson and Smith, 1970).

The situation in the rift zones is entirely different. Rift zones seem to be the zones where new tectonic plates are created by upwelling of mantle material: volcanism from opening fissures is therefore the natural consequence of this process. The volcanoes originating at the oceanic rifts on top of the mid-ocean ridges may be transported with the spreading tectonic plates; thus the observed average relationship between the size of oceanic volcanoes and their distance from the axis of the ridges may be explained simply as growth with age of drifting volcanoes (Menard, 1969).

3.3. Mechanics of volcanic eruptions

3.31. General remarks

The mechanics of a volcanic eruption is a complicated process. It is not our task here to review all the theories of magma formation, etc., that have been advanced in the literature, but rather to present the estimates of energies liberated and the possible mechanisms of the apparent phenomena.

Thus, we shall first review the attempts that have been made to assign quantitatively an intensity or magnitude to an eruption. We shall then review the attempts which were made to comprehend the driving forces in such eruptions.

3.32. Intensity, energy and magnitude of eruptions

In order to describe the strength of volcanic eruptions, it is necessary to have a means of classifying their effects according to

TABLE V

Intensity, volume, magnitude and energy of eruptions

Intensity	Volume (km^3)	Class A		Class B	
		magnitude	energy (erg)	magnitude	energy (erg)
I	10^{-5}	5.8	$2\cdot10^{20}$	5.3	$3\cdot10^{19}$
II	10^{-4}	6.4	$2\cdot10^{21}$	5.9	$3\cdot10^{20}$
III	10^{-3}	7.0	$2\cdot10^{22}$	6.6	$3\cdot10^{21}$
IV	10^{-2}	7.6	10^{23}	7.2	$3\cdot10^{22}$
V	10^{-1}	8.2	10^{24}	7.8	$3\cdot10^{23}$
VI	1	8.8	10^{25}	8.4	$3\cdot10^{24}$
VII	10	9.4	10^{26}	9.1	$3\cdot10^{25}$
VIII	100	10.0	10^{27}	9.7	$3\cdot10^{26}$
IX	>100	10.6	$9\cdot10^{27}$	10.3	$3\cdot10^{27}$

some quantitative intensity scale. The first to devise such a scale appears to have been Tsuya (1955), who based his scale solely on the volume of material ejected. Tsuya's intensity scale is shown in Table V.

If an attempt is made to correlate the volume of the ejecta with the energy of an eruption, one should make a distinction between the ejected material consisting chiefly of molten lava (type A) and ejected material consisting of solid or old material (type B): the reason for this distinction is that molten lava contains far more energy than solid material (Hédervári, 1963).

The energy liberated in a volcanic eruption can be estimated by a phenomenological investigation of the various effects produced by it (Yokoyama, 1956, 1957a; Imbo, 1965). Thus, energy is released as kinetic energy in the ejection of pyroclastic materials from the crater, as thermal energy in the form of the heat contained in the outpouring lava and gasses, as seismic energy in the volcanic earthquakes, and as energy used in the breaking up of rock material, etc. It can be shown that all forms of energy released can be neglected in comparison with the thermal energy. The latter is determined by the mass of ejected material and by its temperature. The pertinent formula is:

$$E = V\rho(\Delta T\cdot c + L) \qquad (3.32\text{-}1)$$

where E is the thermal energy liberated, V the volume of the ejected material, ρ its mean density, ΔT the temperature (°C) of the lava above air temperature, c the specific heat of the lava (0.25 cal./°C), and L the heat of fusion ($L = 0$ for solid material, $L = 50$ cal./g for molten lava). The above formula yields E in calories; a conversion factor must be used if other units are used.

Since the density and temperature of the ejected material is fairly constant, the energy output depends essentially on the volume V, provided the two classes of eruptions with $L = 0$ and $L = 50$ cal./g (solid or molten material) are distinguished.

It is possible to introduce a "magnitude" M of an eruption by a method similar to that used for earthquakes. The pertinent relation corresponding to eq. 2.24-3 is:

$$\log E = \alpha + \beta M \tag{3.32-2}$$

where the values for α and β vary with each author. Hédervári (1963) chose $\alpha = 11$, $\beta = 1.6$ for the definition of the eruption magnitude M. Perusing values for many eruptions, he then obtained the following relations between the volume V (m^3) of the ejecta and the magnitude M.

For class A eruptions:

$$M = \frac{\log V + 6.08}{1.67} \tag{3.32-3}$$

For class B eruptions:

$$M = \frac{\log V + 4.95}{1.59} \tag{3.32-4}$$

Hédervári (1963) applied the last two formulas to a large number of volcanic eruptions. He derived, in this fashion, a correlation between Tsuya intensities and magnitudes for class A and B eruptions. Hédervári's correlation is shown in Table V. The corresponding energies can easily be calculated from eq. 3.32-2; they are also shown in Table V.

3.33. The driving force in an eruption

The driving force needed for the ascent of the magma and its final violent eruption must be sought in phase transitions of multi-component melts. The simplest of these melts consists of water (-vapour) "dissolved" in some volcanic glass. The relationship between the quantity c of a gas in solution (expressed as a fraction of 1) at pressure p at a temperature of 1,000°C is (Rittmann, 1961):

$$c = \frac{ap}{b + p} \tag{3.33-1}$$

where a = 11.4 and b = 1.08 megabar for albite.

Assuming that equilibrium exists in a magma column, a change in the external conditions may upset this equilibrium. If this change is in the direction of a lowering of the external pressure, gas may be liberated which leads to an additional decrease of pressure increasing this disequilibrium. This may ultimately lead to an eruption. It is, thus, the expanding gas which causes the violent effects. In fact, it has been found (Richards, 1965) that there is a linear relationship between the logarithms of the gas pressure p and of the kinetic energy E of volcanic explosions. It is of the type:

$$E = 10^9\, p^{4.2} \tag{3.33-2}$$

where E is expressed in ergs and p in bars.

An actual calculation of the eruption phenomenon is a very difficult undertaking, because the changes in viscosity of the melt have to be taken into account. So far it has not yet been achieved.

3.4. Effects of volcanic eruptions

3.41. Introduction

A volcanic eruption is attended by a variety of catastrophic effects. The molten rock flows from the crater as lava; if a volcanic cone has already been built up by previous eruptions, veritable rivers of lava may be formed which engulf everything in their way until they become solidified upon cooling.

Lighter materials, i.e., rock materials which are mixed with gases, may flow as a hot avalanche from the vent; this leads to nuées ardentes and ash flows. If the rock is even more fragmented, it may be carried in the atmosphere as ash for a long distance and will lead to an ash fall.

Finally, an eruption causes not only the deposition of material on the surface of the earth, but also a withdrawal of a corresponding amount of material from its interior. The cavity which is thus created may collapse leading to a feature which is called a caldera. This collapse may have catastrophic effects.

The phenomena mentioned will be dealt with later in this book.

3.42. Dynamics of a lava stream

The hot lava discharged from a volcano may flow over very large distances. Particularly in prehistoric times, incredibly large sheets of lava have been laid, such as on the Columbia plateau of Washington State (Waters, 1955). It seems difficult to understand the dynamics of such large flows in view of the viscosities that are generally believed to apply to lavas.

The dynamics of a lava flow may be described in a first approximation as that of a Newtonian liquid. A more complicated linear model, however, may actually be more adequate. Many measurements of the viscosity of volcanic rocks under various conditions have been reported by Murase (1962); some semi-empirical methods to predict viscosities of magmas from their composition have been published by Bottinga and Weill (1972) and by Shaw (1972).

During the occurrence of a lava flow, the flowing material cools down which leads to an increase in viscosity. The viscosity change with temperature must be taken into account in an attempt to make a dynamical model of a lava flow. The relationship between viscosity η and absolute temperature T is usually taken as:

$$\eta = \eta_0 \exp(b/T) \tag{3.42-1}$$

where η_0 and b are constants.

This relation follows from general considerations of absolute reaction rates (Glasstone, Laidler and Eyring, 1941). For basalts values of η_0 range between 1.3 and $6.0 \cdot 10^{-6}$ Poise and values of b range between 2.65 and $2.73 \cdot 10^4$ °K (Daneš, 1972). The melting-point temperatures are around 1,150°. The values for andesitic lavas are different, inasmuch as they are generally more viscous and are extruded at a lower temperature (Sakuma and Nagata, 1957). The viscosity of fresh lavas is less than that of remelts; perhaps volatile components are lost or additional bonds grow in the volcanic glasses with time.

Daneš (1972) has developed a simple model for a lava flow, assuming purely Newtonian behaviour. In an inclined layer of viscous fluid in laminar flow, the equation of motion is:

$$\frac{\mathrm{D}u}{\mathrm{D}t} = -\frac{1}{\rho}\frac{\partial p}{\partial x} + g \sin \beta + \frac{\eta}{\rho}\frac{\partial^2 u}{\partial y^2} \tag{3.42-2}$$

where x is the coordinate parallel to the flow, y is the coordinate normal to it, u is the flow velocity, ρ the density and η the viscosity of the fluid, p the pressure, g the gravity acceleration and β is the slope angle of the bottom of the layer. For a layer of thickness h, one has:

$$p = \rho g h \cos \beta \tag{3.42-3}$$

If β is small, one can set $\sin \beta \sim \beta$ and $\cos \beta \sim 1$. One has then for steady-state conditions:

$$\rho g \left(-\frac{\partial h}{\partial x} + \beta \right) = -\eta \frac{\partial^2 u}{\partial y^2} \tag{3.42-4}$$

A further approximation can be made, if $\partial h/\partial x$ is assumed to be constant equal to $-\alpha$. Eq. 3.42-4 can then be integrated from y = 0 to $y = h$; one obtains for the boundary conditions $u = 0$ where y = 0 and $\partial u/\partial y = 0$ when $y = h$:

$$u = \frac{1}{2\eta} \rho g (\alpha + \beta)(2h - y) y \tag{3.42-5}$$

One notes:

$$u_{max} = \frac{1}{2\eta} \rho g(\alpha + \beta) h^2$$
$$\langle u \rangle_{ave} = \tfrac{2}{3} u_{max} \qquad (3.42\text{-}6)$$

Whilst a glowing lava flow proceeds downhill, it cools down. The cooling is primarily described by the law of Stefan and Boltzmann:

$$-\frac{dT}{dt} = \frac{\epsilon\sigma}{\rho L a h} T^4 \qquad (3.42\text{-}7)$$

where ϵ is the specific emissivity of the lava, σ the universal Stefan-Boltzmann constant, L the energy-heat equivalent, a the specific heat capacity of the lava and T the absolute temperature. In practice, information about the duration of a lava flow is hard to obtain; therefore the time-dependence is replaced by a dependence on x:

$$dt = dx / \langle u \rangle_{ave} \qquad (3.42\text{-}8)$$

which yields:

$$\frac{dT}{T^4 \exp(b/T)} = \frac{3\epsilon\sigma\eta_0}{\rho^2 \eta a h^3 g(\alpha + \beta)} \qquad (3.42\text{-}9)$$

or, after integration:

$$\exp(-b/T) \left[\left(\frac{b}{T} + 1 \right)^2 + 1 \right]_{T_0}^{T_s} = \frac{3\epsilon\sigma\eta_0 b^3 x}{\rho^2 \eta a h^3 g(\alpha + \beta)} \qquad (3.42\text{-}10)$$

Here, T_0 is the initial temperature and T_s the temperature at the end (solidification) of the lava flow (1,150°C). In nature, flows with $x = 300$ km have been found. For $\epsilon = 1$, $\eta_0 = 1.318 \cdot 10^{-6}$ Poise, $b = 2.65 \cdot 10^4$K, Daneš found minimum thicknesses and minimum temperatures to produce such a lava flow as shown in

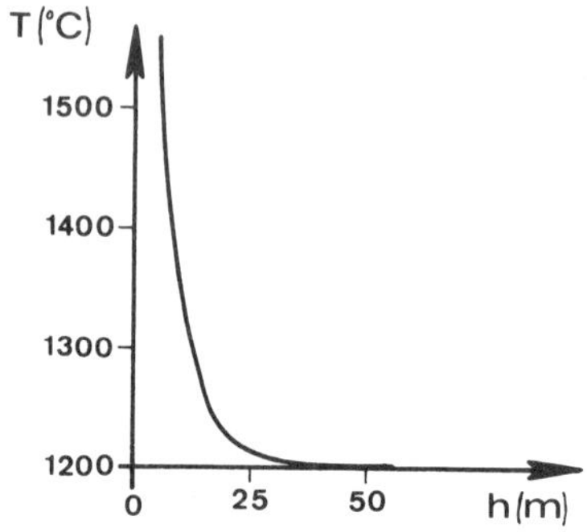

Fig. 9. Minimum thickness and temperature to produce a 300-km lava flow, if $\alpha + \beta = 0.001$. (After Daneš, 1972.)

Fig. 9. It is seen that thin flows (25–50 m thick) can travel very large distances. This explains the long reach of lava flows.

The above theory considers the lava stream as a homogeneous viscous fluid. In fact, for such a stream cooling takes place from the surface inward which leads to the formation of a cold "skin". Then, the lava stream advances locally when protrusion of bulbous fingers from the lava sack breaks its skin (Sakuma and Nagata, 1957). Furthermore, non-Newtonian behaviour becomes pronounced upon cooling (Shaw, 1969). It is, of course, difficult to make exact theoretical calculations of such a phenomenon. Nevertheless, the artificial breaking of the skin (by dropping explosives from aircraft) in topographically propitious locations has made it possible to deflect a lava stream, as was the case with a lava stream from Mt. Mauna Loa in 1942 which endangered the city of Hilo (Rittmann, 1960, p. 63). Depending on the viscosity of the lava, even low earth levies which have been quickly thrown against a lava stream may help to save plantations from destruction. A large-scale attempt to control a lava stream which endangered a city was made by cooling the lava with water, in the spring of 1973 during an eruption on the island of Heimaey in Iceland (Williams and Moore, 1973). Sea water (up to 1 m^3/sec) was pumped from barges to various parts of the lava flow which had an original temperature of about 1,030°–1,080°C. It was found that the cooling achieved by the water had a definite effect on the flow of the stream which was greatly slowed down. First, the margin of the lava flow was cooled, and then the water was applied to progressively higher areas. It was found that it took about one day for a discharge of 200 l/sec at any one point to become effective; after about 2 weeks of continuous application at the point steaming would stop, indicating that the lava had been cooled to below 100°C at the point of discharge. The cooling procedure used in Iceland constitutes one of the most successful attempts at controlling a lava flow to date.

3.43. Ash flows

The ash-like ejecta of volcanic eruptions are transported and deposited in two different mechanical ways: volcanic debris which is blown high in the air may be deposited on either land or in water and is called an *ash fall*; it contrasts with an *ash flow*, material that accumulates from a hot, incandescent turbulent mixture of debris and gas that travels rapidly down the slope of a volcano. Gravity supplies the driving force that impells an ash flow downhill, whereas in an ash fall there is only horizontal wind drift and decay of turbulence, gravity simply withdrawing material from the debris cloud as it moves downwind. We shall treat in this section the mechanics of ash flows, and in the next section (3.44) the mechanics of ash falls.

A classic example of an ash flow in the form of a glowing cloud (nuée ardente) occurred during the eruption of Mt. Pelée and was described by Lacroix (1904). The possible mechanical schemes of such flows were discussed by Fisher (1966), Shreve (1968b) and Allen (1971a).

Fisher (1966) used the boundary layer concept of the fluid mechanics, assuming that a low-viscosity layer develops between the actual flow and the ground. Shreve (1968b) assumed that such a lubricated layer would be provided by trapped and compressed air, a model which he also proposed for catastrophic landslides (cf. Sect.4.74). Allen (1971) assumed that a nuée ardente represents a case of a turbidity current in air. Unfortunately, the mechanics of turbidity currents (cf. Sect.6.52) is only imperfectly understood, so that it would be difficult to make predictions about the reach, speed, etc., of a possible nuée ardente in an expected volcanic eruption.

3.44. Ash falls

In an ash fall, the ejecta from a volcanic eruption are carried in turbulent suspension for a long distance in the atmosphere. As the turbulence decays, the carrying capacity of the air decreases so that the ash gradually falls to the surface of the earth. Ash falls are the products of aeolian differentiation in which particles are segregated by their fall velocity as they are transported downwind.

Reviews of the available observations on ash falls have been provided by Fisher (1964) and by Eaton (1964). Scheidegger and Potter (1968) have studied the fall-out mechanism based on a physical model. Accordingly, the following assumptions have to be made:

(1) In an ash fall, a certain volume of ash is ejected into the air during a time interval T with an initial distribution of particle sizes and densities given as a power function of the settling velocity.

(2) The ash is suspended in turbulent motion in the air; the turbulence decays according to standard decay laws.

(3) The carrying capacity of the turbulent air is analogous to that in a river so that with decay of turbulence there is fall-out of ash. The fall-out time of ash at any downwind point is proportional to the ash-eruption time.

(4) Superimposed upon the above effects is a constant mean wind away from the source of the ash.

The above model is analogous to the one employed by Scheidegger and Potter (1965) in a general discussion about graded bedding. Here, however, the downwind transport caused by a wind with mean velocity v is considered. The final result sought is an analytical expression for the patterns of particle size δ and thickness h of an ash fall at various distances x from its source. One obtains:

$$\delta = \text{const}\, (v/x)^{m/2} \qquad (3.44\text{-}1)$$

$$h = \text{const}\; T(v/x)^{\beta m+1} \qquad (3.44\text{-}2)$$

where T is the duration of the ash eruption, m is the turbulence-decay constant (ranging between 1 and 5/2) and β is a constant depending on the initial particle-size distribution. The formulas show a decrease of particle size δ and ash-fall thickness h with distance x from the source. An attempt to check the formulas with existing observations yielded a satisfactory result.

3.45. Caldera formation

When the lava has been ejected from a volcanic vent by gas expansion, a hollow space is left behind. This hollow space may subsequently collapse under the the weight of the overburden, leading to the appearance of a depression on the earth's surface, which is called a *caldera*.

Whether a caldera forms or not is a question of the stability of underground cavities. The problem has been discussed by Scheidegger (1970a, see p.414ff.) in connection with the genesis of caves. It appears that the stability of a cavity is independent of the size of the cave. In a stress state, which is a pure (isotropic) compression p far away from the cavity (overburden pressure), the maximum shearing stress at the wall of a spherical cave is:

$$\sigma_{\text{shear max}} = \frac{3}{2}p \tag{3.45-1}$$

A comparison of this shearing stress with the shearing strength of the rock material will indicate whether the cavity will collapse or not.

The formulas for a triaxial stress state are much more complicated. The reader is referred to the cited literature for the details.

When the cavity collapses, subsidence spreads to the surface. Denoting the vertical coordinate by z, the horizontal ones by x and y, then the subsidence w of a layer at height z above the cavity obeys the equation:

$$\frac{\partial w}{\partial z} = K\left(\frac{\partial^2 w}{\partial x^2} + \frac{\partial^2 w}{\partial y^2}\right) \tag{3.45-2}$$

which is a diffusivity equation (Litwiniszyn, 1963). Scheidegger (1966) has given a general explanation based on statistical mechanics for the provenance of this diffusivity equation.

For practical applications, the displacement at the level of the cavity ($w = w(x, y, z = 0)$) must be given, then the solution of the diffusivity equation (3.45-2) yields the form of the caldera at the surface. The constant K must be estimated from field observations.

A completely different view of the genesis of a caldera was taken by Oliveri del Castillo and Quagliarello (1970) who assumed that when the pore water was heated by ascending magma it caused the pore pressure to rise until the rock would split internally. This would allow pore fluid to escape leading to a subsequent reduction in pore pressure which would cause caldera-like subsidence regions in the surrounding area.

3.46. Volcanic bombs

The explosions which may accompany a volcanic eruption can cause the ejection of volcanic "bombs".

Such bombs are simply stones hurled a long distance. Because of their size (their diameter is often larger than 50 cm) they follow generally parabolic trajectories; air resistance is only of minor importance. The energy of a volcanic eruption transformed into the kinetic energy of the bombs is only a minor fraction of the total energy released, as has been noted above.

For a given initial velocity v_0 the maximum reach D of a bomb is attained if the latter is shot out at an angle of 45°. It is:

$$D = v_0^2/g \qquad (3.46\text{-}1)$$

Thus, the initial velocity v_0 can be estimated from the farthest reach D of volcanic bombs caused by an eruption. In this fashion, initial velocities have been estimated which lie often above 170 m/sec (Yokoyama, 1956, 1957a, b; Sakuma and Nagata, 1957). With this velocity, the maximum reach of a bomb is about 3 km. Gorshkov (1959) obtained v_0-values for bombs in the giant eruption of Mt. Bezymianny on Kamchatka in 1956 of 500–600 m/sec; in that case, bombs reached distances of 25 km and more.

Needless to say, it is difficult to design an effective protection against missiles such as volcanic bombs. The event has to be forecast in advance and the areas likely to be affected have to be evacuated.

3.5. Statistics and prediction of volcanic eruptions

3.51. The problem

Volcanic eruptions represent catastrophes that can cause great damage to structures as well as loss of life. As with the occurrence of earthquakes, there is a great need for predicting the time and severity of impending eruptions in order to forewarn the affected population in time. Similarly, an evaluation of the risk to structures presented by the possibility of a volcanic eruption is important for insurance purposes.

The problems indicated above have not yet been completely solved, but the many attempts which have been made will be discussed in the following sections.

3.52. Statistics of eruptions

If one considers a volcano in the course of time, one notes that eruptions are events of relatively short duration. The repose periods between eruptions can therefore be treated as a stochastic series and its characteristics can be investigated. This type of approach has been taken by Wickman (1966) who discussed the general principles and applied them to volcano records from various parts of the world.

The basis of Wickman's treatment of volcanic records lies in the introduction of two functions. The first is the survivor function $F(x)$ which gives the probability that a repose period has not ended up to the time interval x since the last eruption. The second is the age-specific eruption rate $\Phi(x)$ which is connected to $F(x)$ as follows:

$$\Phi(x) = -\frac{\mathrm{d}}{\mathrm{d}x}[\ln F(x)] \qquad (3.52\text{-}1)$$

It gives the probability density of the occurrence of an eruption during unit time after the time x has passed since the last eruption.

With the above concepts, each volcano can be treated individually. In practice, the survivor function F is replaced by a survivor number N:

$$N = N_0 \, F(x) \tag{3.52-2}$$

where N_0 is the total number of repose periods on the volcanic record, $N(x)$ is the number of periods that outlasted the time interval x. Plotting $N(x)$ yields a step function which must be smoothed (by eye) and scaled by N_0 to yield $F(x)$. Then, differentiation of the logarithm of this function gives $-\Phi(x)$. The latter gives the probability of eruption in a time unit. As an example, we show in Fig.10 the survivor functions and the eruption rate of Mt. Bromo in Java.

Wickman (1966) used the above procedure for the evaluation of volcanic records and for the estimation of the probability of an eruption in the near future, which is valuable for insurance purposes.

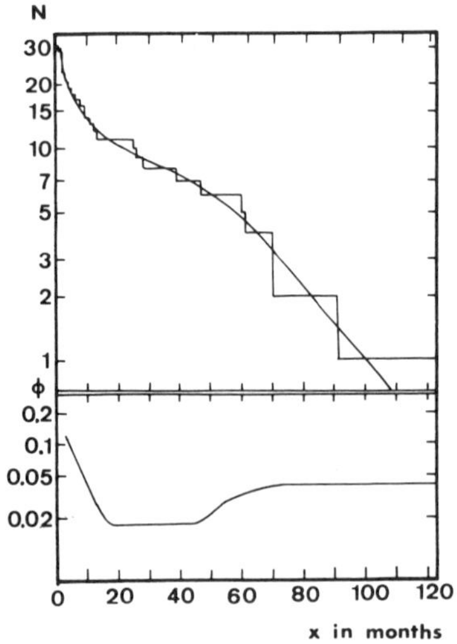

Fig. 10. Repose periods (number N of periods longer than time x) of Mt. Bromo (top); eruption rate as calculated therefrom (bottom). Time unit: month. (After Wickman, 1966.)

The procedure indicated above is entirely empirical. An analysis of the eruption rates showed that some volcanoes are simple Poisson volcanoes, i.e., the eruption rates Φ are constant. The probability of n eruptions within a given period of time t is then:

$$P_n = \frac{(\Phi t)^n}{n!} \exp(-\Phi t) \qquad (3.52\text{-}3)$$

where Φ is the parameter of the Poisson distribution. The volcanoes Etna, Kelud, Mauna Loa, Merapi and Popocatepetl are of this type. Other volcanoes may fit into this category if allowance is made for the statistical deviations in sampling; these are due to the shortness of the records. Eruption rates for these volcanoes lie between $50 \cdot 10^{-3}$ month^{-1} and $0.5 \cdot 10^{-3}$ month^{-1}. Thus, Φ varies by a factor of about 100.

Many volcanoes do not fit into a simple Poisson process. There may be several magma chambers, or there may be a specific loading time. Thus, the stochastic behaviour of each volcano has to be studied individually to enable one to make specific predictions.

3.53. Attempts at predictions

Attempts at predictions of volcanic eruptions have been based on the surveillance of known volcanoes. A variety of phenomena have been found to be forerunners of violent eruptions, although no sure means have been devised as yet to make completely reliable predictions. The various methods by which prediction has been attempted have recently been summarized by Unesco (1971).

Accordingly, physical and chemical methods have been used to search for precursors of volcanic eruptions. Premonitory earthquakes have been found in many instances. A characteristic peak in earthquake frequency was found one day before the eruption of Mt. Usu (Unesco, 1971, p.25). Geodetic (notably tilt-meter) measurements have been particularly successful in monitoring the activity of Kilauea Iki whose 1959 eruption was accurately forecast. This was possible because the lava ascent in this volcano occurs very rapidly (50 km in 6 months) causing an inflation of the whole volcano which disappears again after every eruption.

A physical model of this phenomenon has been provided by Mogi (1958) who obtained the following horizontal deformation Δd on the surface at the distance d from a point above the centre of a small sphere of radius a with internal pressure P buried at depth f:

$$\Delta d = \frac{3a^3 P}{4\mu} \frac{d}{(f^2 + d^2)^{3/2}} \tag{3.53-1}$$

where μ is the shear modulus of the material. An alternative model was proposed by Yokoyama (1971).

No similar effects can be observed, however, in volcanoes which are characterized by explosive eruptions.

Magnetic and electrical surveillance of volcanoes yielded correlations between anomalous changes in the magnetic and electrical fields and volcanic activity (Yokoyama, 1969). The changes in the magnetic fields could be due to temperature (the Curie-point is exceeded) and stress (magnetostriction) effects. The correlations, however, are so far entirely observational and not systematic: one simply notes changes during volcanic eruptions. The same is the case for changes in resistivity. Reliably predictive statements, therefore, cannot be extracted from such measurements as yet.

An increase in thermal activity (fumaroles) has sometimes also been observed prior to volcanic eruptions, but need not always precede a catastrophic event. A general correlation has been claimed to exist between the instant of a volcanic eruption and the maximum fortnightly tidal amplitudes (Mauk and Johnston, 1973).

Finally, changes in the chemical composition of emitted gasses have been linked with impending eruptions.

None of the features mentioned above are reliably predictive, with the notable exemption of the horizontal movements on Kilauea by which, as mentioned, the 1959 eruption of the Iki crater could be accurately predicted. In this case, however, conditions were particularly favorable inasmuch as the ascent of the magma could practically be traced. In more explosive eruptions, such conditions cannot be expected to prevail. The seismic method of counting the frequency of small tremors gives generally the most accurate indications of an impending eruption, which may be heralded by an increase in the normal frequency pattern. However, although eruptions are generally connected with an increase in the number of earthquakes per unit time, such an increase is not always followed by an eruption, which fact can lead to many "false alarms". Conversely, if it is found that earthquake swarms felt by the population of an area have epicentres far from the location of a suspect volcano, it is possible to allay the fear that these earthquakes might herald an eruption.

3.54. Volcanic zoning

In the vicinity of a volcano it is possible to zone areas most likely to be affected by the effects of further outbursts. These "zoning maps" are based on past experience and on an estimation of likely events from onsite inspection. The areas likely to be affected by nuées ardentes, bombs and lava streams are delineated. The probability of an eruption can be estimated by the stochastic methods outlined in Sect.3.52, so that fairly precise estimates of the risk to structures can be obtained. People can be forewarned to leave endangered areas.

This type of approach was used by Neumann van Padang (1960), by Healy (1963) and by Zen and Hadikusumo (1965), as well as by others.

3.6. Other volcanic effects

3.61. Description

There is a variety of phenomena which are associated with volcanic activity that do not fall into the categories which have already been discussed.

The craters of volcanoes commonly become filled with water. It is possible that such crater lakes become unstable and will empty suddenly, causing a large flood wave and mud flow. These flows are called *lahars*.

Volcanic activity often engenders hydrothermal effects. Geyser activity and explosion craters may be the result.

We shall discuss these phenomena below.

3.62. Lahars

Lahars are mud flows which originate in volcanic craters that have been filled with water. When a crater lake empties, it may be the result of the collapse of a sector of a volcano, of subaerial erosion, or of the collapse of an ice barrier on a high-altitude crater.

A disaster caused by a collapse of the last type occurred on 24 December 1953 near Tangiwai in New Zealand: a lahar which originated in the crater lake of Mt. Ruapehu swept away one of the piers of a railway bridge a few minutes before a passenger train was to cross it: five railway cars plunged into the river and 151 people were killed (Healy, 1954). In this case, the failure was probably caused by crevassing in an ice barrier atop Mt. Ruapehu. The prediction of such events must start with a survey of potentially dangerous crater lakes, leading to an evaluation of the stability of the barriers and a calculation of the potential danger of a flood wave if failure should occur (cf. Sect.6). The probability of a lahar developing can also be lowered by constantly draining dangerous crater lakes by a system of tunnels. This was done successfully on Mt. Kelut in Java (Zen and Hadikusumo, 1965). After an eruption, the tunnels may have to be rebuilt.

3.63. Hydrothermal effects

Catastrophic phenomena can also be produced by hydrothermal effects. Muffler et al. (1971) have shown that explosion-type phenomena can be produced if water contained in near-surface rocks suddenly flashes to steam. These explosions are due to the same mechanism as geyser eruptions but are much more violent. They can produce craters which are up to 1.5 km in diameter.

Geyser eruptions are due to the instability existing in a water column which is everywhere at the boiling point. If some water is removed from its top, the pressure throughout the column is reduced, leading to vaporization of some water which, in turn, leads to a further reduction in pressure. This causes more water to be vaporized so that the geyser erupts.

The same chain reaction occurs in hydrothermal explosions.

However, the steam ejects much of the surrounding rock so that no periodicity as in the case of geysers can develop: a single crater is formed in a single eruption.

The triggering mechanism is assumed to be some event causing a sudden decrease in hydrostatic pressure, which enables a large quantity of hot water below to flash to steam. Such events may be the draining of a lake above, the formation of fissures due to an earthquake, etc. It is known that earthquakes may affect the hydrothermal regime of an area (Ward, 1972).

Muffler et al. (1971) made calculations of the energy which is released when superheated water in a hot-spring system flashes to steam, and compared it with the energy that is required to form a crater of a given size. They found reasonable agreement, particularly for the larger craters (1–1.5 km in diameter), so that the envisaged mechanism has been substantiated.

Chapter 4. ACCIDENTS ON SLOPES

4.1. General remarks

4.11. Introduction

Potentially any slope can be the cause of a catastrophic event. Owing to the action of gravity and lack of stability, the slope can collapse with dire consequences. Much work has already been accumulated in the engineering literature on stability criteria; however, this is not sufficient for an understanding of the catastrophic events themselves. What is needed is a proper elucidation of the actual dynamics of the events in question. Needless to say, this is an extremely difficult undertaking.

The catastrophic events that can take place on slopes need not necessarily be rapid. The slow but inexorable sliding over a year or so of a mountainside, including the buildings thereon, into the sea is perhaps only slightly less of a catastrophe than more violent displacements. Therefore, creep-type phenomena will have to be investigated as well as the more obvious "slides" and "falls" of soil and rock.

The subjects of this chapter will be presented as follows. First (4.1) the general rheology of soil and rock will be described. Then (4.2) the basic stability conditions for existing slopes will be discussed. Much literature has been published on this problem.

Subsequently, the various slope motions will be analysed. It is difficult to set up a general mechanical scheme for them, because the various phenomena represent characteristic combinations of different mechanical processes. Thus, the procedure of Nemcok et al. (1972) of dividing the slope-motions into creeping motions, flows, slides and falls is, in our opinion, untenable in the view of the phenomenological facts. Is a rock glacier creeping, sliding, flowing or falling? In fact, it does all these things.

The best one can do is to make an inventory of the various possible events and to group them into characteristic classes. First of all, there are pure rock-mechanical phenomena (Sect.4.3). These include events produced by rock stresses as well as rock falls and exfoliation.

A further rather well-defined group of events is connected with the mechanics of scree and talus accumulations (Sect.4.4). This group includes débris avalanches, the theory of single rolling stones and the general dynamics of scree slopes.

Events connected with surficial movements of slopes form the next class of phenomena (Sect.4.5). This class includes skin creep, which involves the zone of the seasonal temperature and moisture fluctuations; it rarely affects a depth of more than 3 m. Solifluction is a special type of skin creep under niveal conditions.

When the motion is deeper-lying, one has the phenomenon of deep mass movements (Sect.4.6). This encompasses creeping and sliding motions in soil and rock including the slow closure of valleys.

The various kinds of landslides are very important (Sect.4.7). No single mechanical scheme exists for all landslides, since there are many different types. In view of the importance of the phenomenon, the corresponding section will be a large one.

The variety of phenomena in the last section (Sect.4.8) of this chapter can, perhaps, be best characterized by a fairly rapid flowing or slipping mechanism. Earth flows, mud flows, rock glaciers and "muren" belong in this category.

4.12. Soil and rock

The composition of the materials of slopes show two entirely different types of behaviour. Sometimes, the dynamics of a slope can be treated by the classical methods of "soil mechanics", at other times the newer approaches of "rock mechanics" have to be used. Although one has a certain intuitive feeling about what is "soil" and what is "rock", a proper mechanical classification is nevertheless required.

However, such a classification cannot be given in absolute terms. Rocks as well as soils are inhomogeneous substances; what is important is the size of the individual constituents of the substances in relation to the scale of the mechanical phenomena.

The constituents of soils are represented by the individual grains and soil particles. These are ordinarily very small compared with the scale of the displacements under consideration. However, the size of the constituents is much larger in most rock masses than in soils. They are represented by the volumes bounded by the basic joint systems which are present in every rock mass. The resulting cleavage bodies are generally comparable in size with the scales of the mechanical phenomena in which we are interested.

Generally, the procedures of "soil mechanics" are applicable if the size of the constituents in the substance under consideration is much less than the scale of the mechanical phenomena in which we are interested. This is generally the case in soils and in some rock masses. In this case, the body can be considered as "homogeneously inhomogeneous"; one can average statistically over the individual inhomogeneities and express the mechanical properties of the material by macroscopic constants. These, then, represent "equivalent properties" (Wardle and Gerrard, 1972) of the material.

On the other hand, if the constitutive inhomogeneities are comparable in scale with the geometrical scale of the phenomenon in question so that the individual inhomogeneities must be accounted for in the dynamic scheme describing the event, one must apply the methods of rock mechanics. It is seen, thus, that the answer to the question as to whether soil or rock mechanics has to be applied in a particular instance, may not only depend on the material alone, but also on the scale of the phenomenon that is to be investigated (Müller, 1969).

The joints may be classified phenomenologically as follows (Farran and Thénoz, 1965).

(i) Microfissures which are less than 1 micron in width and of very small areal extent, the latter being comparable with the space taken by several molecules.

(ii) Microfractures whose width is up to about 0.1 mm. Their areal extent may be quite considerable, up to 1 m^2.

(iii) Macrofractures which are wider than 0.1 mm and may be up to several m^2 in areal extent. The filling material is of significance with regard to strength properties.

(iv) Faults which are large (on the day-surface up to several thousand kilometers long) geological discontinuities.

The jointedness of rocks is measured by statistical means. One can define a series of indicative parameters. The degree of jointing, k, is represented by the number of intersections of a set of joints per length unit normal to their geological plane; conversely, $1/k$ is the average spacing of the joints of the set under consideration. The unit block is the smallest homogeneous rock volume that is determined by three sets (a, b, c) of joints; its average volume V_{UB} is:

$$V_{UB} = \frac{1}{k_a} \frac{1}{k_b} \frac{1}{k_c} \tag{4.12-1}$$

The extent of joints is given by a further parameter $k^{(2)}$. It denotes the fraction of the joint planes which is actually jointed.

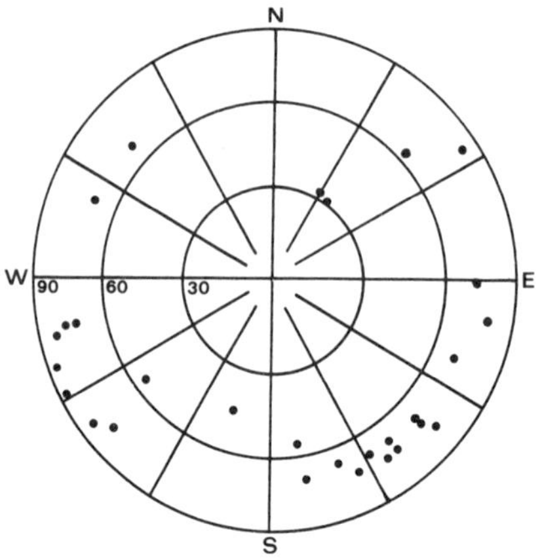

Fig. 11. Example of a diagram showing joints measured in the field. The "dots" represent normal directions of the joint surfaces.

The orientation of joints can be represented as a diagram of observed dip directions on an equal-area representation of a sphere (Fig.11).

4.13. Rheology of soil and rock: elastic range

If small stress changes act upon slope-materials, these materials generally respond in an elastic fashion. The term "elasticity" implies a linear, reversible relationship between the stresses and the strains. For isotropic bodies, this relationship is of the form:

$$\tau_{ik} = \lambda \delta_{ik} \epsilon_{jj} + 2\mu\epsilon_{ik} \tag{4.13-1}$$

where τ_{ik} denotes the stress tensor applied, and ϵ_{ik} the resulting strain tensor. Furthermore, δ_{ik} is the Kronecker tensor and the summation convention has been applied. The constants of proportionality λ and μ are characteristics of the material in question; they are called Lamé's constants.

As is seen, *two* constants are needed for the description of the elastic behaviour of an *isotropic* elastic body. The preference in pure science is for using Lamé's constants as defined above; in engineering, two other constants, called Young's modulus E and Poisson's number m are preferred to λ and μ. The connections between the various elastic constants are given by the following relations:

$$E = \frac{\mu(3\lambda + 2\mu)}{\lambda + \mu} \tag{4.13-2}$$

$$m = \frac{\lambda}{2(\lambda + \mu)} \tag{4.13-3}$$

Earth materials behave elastically as long as the stress changes applied are small enough. For short-period stress cycles (up to a few minutes or so) this is proven by the existence of seismic waves, and for longer stress cycles (up to 12 h or so) by the fact that it is possible to describe the response of the solid earth to the action of tidal forces by means of Love's and Shida's numbers (cf. Melchior, 1966). An elastic response under small loads is also commonly observed in rock samples tested in the laboratory.

The constants λ, μ and E have the dimension of stresses, m is a number. Values of E are commonly around 10^{11} to 10^{12} dynes/

cm^2 for materials found in the earth's crust (Wuerker, 1956; Birch, 1966), *m* is around 0.2. A very extensive table of values of the elastic constants for various earth materials was also given by Müller (1963, p.58 ff.). The equivalent anisotropic elastic properties of layered rock and soil masses have been investigated by Wardle and Gerrard (1972).

4.14. Rheology of soil and rock: creep range

When the load-changes acting upon slope materials are no longer very small, the response is no longer exclusively elastic. It is common for such materials to eventually fracture when the stresses become large enough; but for a considerable time before a fracture occurs, deviations from elasticity are observed. Slow continuous inelastic deformations occurring in a body under the action of a load are denoted by the term "creep".

One has undertaken to set up theoretical models for the explanation of continuous nonelastic responses. The primary response of the material is an elastic one and the task is to describe the observed deviations from it. The first attempt to do this is by making a superposition of a liquid-type of response upon the basic elastic one. Referring symbolically only to one of the off-diagonal elements of the stress and strain tensors, eq.4.13-1 describing elastic behaviour takes the form:

$$\tau = 2\mu\epsilon \qquad (4.14\text{-}1a)$$

where the indices have been omitted. The corresponding equation for viscous liquids is:

$$\tau = 2\eta\dot{\epsilon} \qquad (4.14\text{-}1b)$$

where η is the viscosity.

One can now attempt to construct rheological models by making a superposition of eq.4.14-1a and eq.4.14-1b. This can be done in two ways.

First, one can write:

$$\dot{\epsilon} = \frac{\dot{\tau}}{2\mu} + \frac{\tau}{2\eta} \qquad (4.14\text{-}2)$$

A substance which obeys this equation is called a "Maxwell liquid". Indeed, it exhibits some features of a liquid: the material can only be in equilibrium, at least asymptotically, if no shearing stresses are present; if shearing stresses are impressed upon such a material, but the body is confined so that no strains can develop, the stresses will disappear eventually according to an exponential decay law. The time constant T is equal to:

$$T = \eta/\mu$$

Secondly, eqs.4.14-1a and 4.14-1b can be superposed to yield:

$$\tau = 2\mu\epsilon + 2\eta\dot{\epsilon} \qquad (4.14\text{-}3)$$

A substance obeying this equation is called a "Kelvin body". Its characteristic properties are found if a stress change is applied and a response from the body is observed. It is seen that the body will asymptotically attain the deformation corresponding to an elastic material with the Lamé constant μ. This occurs according to an exponential build-up law; the build-up time T is again given by:

$$T = \eta/\mu$$

The two linear models described above which can be obtained by the superposition of an elastic and a viscous behaviour, do not suffice to accurately describe the responses observed in slope-materials. Therefore, nonlinear responses between stresses and strains must be experimented with. Scheidegger (1970b) has proposed the following equation:

$$\dot{\tau} = 2\eta\ddot{\epsilon} + \beta(\dot{\epsilon} - C)^2 \qquad (4.14\text{-}4)$$

Here, η is again some kind of viscosity, β is a creep factor and C is a constant. Upon integration for a constant stress τ_0 eq.4.14-4 leads to:

$$\epsilon = \frac{\tau_0}{\mu}\,[1 + q\,\ln(1 + Bt)] \qquad (4.14\text{-}5)$$

where μ is the Lamé rigidity and q and B are constants. Eq.4.14-5 was proposed by Lomnitz (1956) upon empirical grounds.

A body which obeys the Lomnitz equation (4.14-5) exhibits creep under constant stress τ_0 in which the strain is a logarithmic function of time. This behaviour is called "logarithmic creep".

In addition to creep laws of the logarithmic type, nonlinear "Glen" type laws (cf. Sect.5.23) have also been applied when describing soil and rock creep.

The various rheological models discussed above are tested, of course, when compared with empirical facts. Of greatest importance are direct investigations of the creep behaviour of real rocks under realistic natural stress conditions. In this connection, laboratory experiments are most significant. Rock samples have been subjected to an isotropic stress state to simulate the overburden effect; shearing stresses have then been superposed. One finds that only few rocks exhibit a linear behaviour corresponding to eq.4.14-2 or 4.14-3. Most rocks are best described by a logarithmic (and not "Glen"-type) creep law of the type of eq.4.14-5. For practical purposes, this equation may be modified to read (Parsons and Hedley, 1966):

$$\epsilon = \frac{2\eta}{\beta} \ln[b(t - t_0) \tag{4.14-6}$$

From experiments, Parsons and Hedley (1966) obtained for the factor $2\eta/\beta$ values ranging from about 5.5×10^{-6} for sandstones to about 6.7×10^{-5} for potash.

It may be mentioned that the logarithmic creep equation does not only correspond to laboratory experiments with rocks, but also seems to explain many geodynamic phenomena (Scheidegger, 1970c, d, 1971, 1972). The damping of seismic waves, the magnitude and frequency patterns in aftershock sequences of large earthquakes (Ranalli and Scheidegger, 1969), the phase shifts in the tides of the solid earth (Jeffreys, 1958) and the characteristics of post-Glacial uplift patterns after the last ice age can only be explained by logarithmic creep. The argument may be summarized as follows.

(i) The damping of seismic waves and of the free oscillations of the earth requires a quality factor Q which is frequency-independent. Linear rheological equations cannot account for this, but a logarithmic-creep equation can.

(ii) The magnitude and frequency patterns of earthquakes in seismic aftershock sequences cannot be explained by a linear strain rebound theory (Ranalli and Scheidegger, 1969). The pattern points toward a nonlinear logarithmic-creep equation of the type of eq.3.14-5.

(iii) The phase shift in the earth tides can be explained by a linear or by a nonlinear rheological model. However, if a linear model is used, the required rheological constants come out as very different from those for exponentially creeping rocks. If a nonlinear creep model is used, no such difficulties arise.

(iv) The Chandler wobble can be explained by either a linear or a nonlinear model, so that this phenomenon leads only to equivocal evidence.

(v) The curves of post-Glacial uplift for areas that were glaciated during the last ice age, do not correspond to viscous rebound. Of the rheological possibilities given at the beginning of this chapter, the uplift curves can only be explained by logarithmic creep.

Thus, for all temporal and spatial scales, the continuous deformation deviations from elasticity can only be explained by assuming that the rheological equation of soils and rocks is that of logarithmic creep.

4.15. Rheology of soil and rock: failure range

As noted above, the deformations in slope-forming materials eventually become discontinuous if the stresses are large enough. When discontinuities appear in a material, the latter is said to undergo "failure".

It is one of the fundamental tasks of practical mechanics to establish the conditions under which a material will fail. These conditions must involve the stresses in some fashion. Moreover, the failure conditions must be independent of the representation of the stress tensor in a particular coordinate system, so that they must depend solely on tensor *invariants*.

A useful tool for the representation of failure conditions in

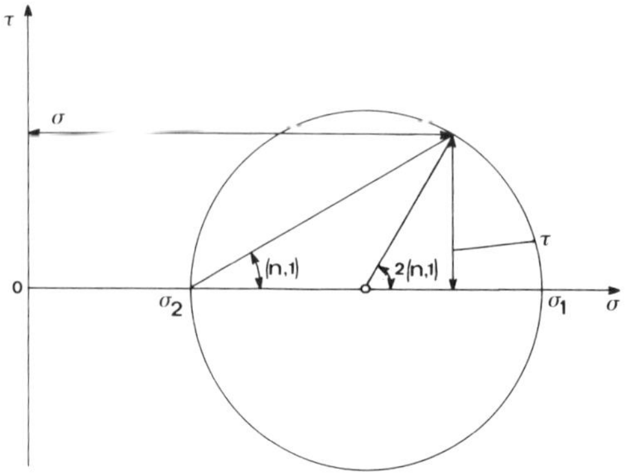

Fig. 12. Mohr circle for a two-dimensional stress state.

two-dimensional stress states is the Mohr (1928) circle as shown in Fig. 12. In this figure, the *1*-direction corresponds to the first principal stress direction: no shear stress is acting on the corresponding surface element and the normal stress is σ_1. The second principal stress direction is the *2*-direction. Here, the normal stress is σ_2 and the shearing stress is again zero on the corresponding surface element. The stresses σ and τ acting on a surface element whose normal n subtends an angle $(n, 1)$ with the *1*-direction are then read off the Mohr diagram as shown in Fig. 12. In the Mohr-circle representation, the yield condition can be indicated by limiting curves which the Mohr circles representing the stress state at any point in the medium may not exceed.

The failure criteria were discovered to be quite different for those materials which can be described by classical soil mechanics and for those which must be described by rock mechanics.

"Soils" consist essentially of assemblages of individual grains and particles. The coherence of such assemblages depends on an inner cohesion and on friction between the grains. The limiting shearing stress τ_s that can be supported is given by the so-called Coulomb equation:

$$\tau_s = \sigma \tan \phi + c \qquad (4.15\text{-}1)$$

where σ is the normal stress, c the cohesion and ϕ the angle of internal friction of the material in question. In thy Mohr diagram,

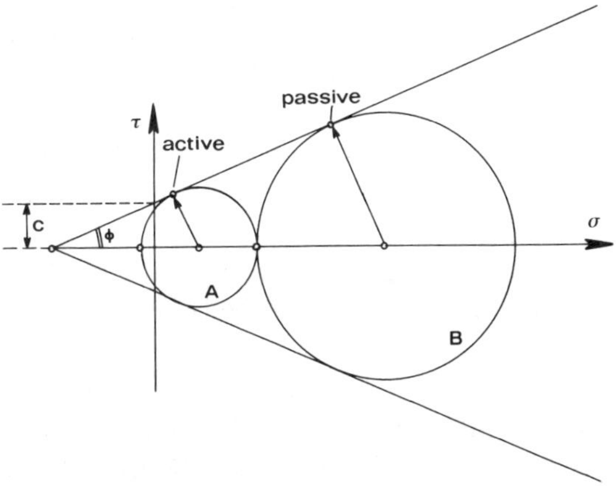

Fig. 13. Active and passive Rankine states.

eq.4.15-1 is represented by two straight lines as shown in Fig.13. If the Mohr circle just touches the limiting shear lines, the material is said to be in a Rankine (1857) state of failure. If the normal stress acting on an element in the *1*-direction is given, then there are two Rankine states of failure possible. They correspond to the Mohr circles *A* and *B* in Fig.13 and are called "active" and "passive" Rankine states, respectively. In Fig.13, the active Rankine state is represented by the circle *A*, the passive one by the circle *B*. The surface elements for which the limiting shears (shear surfaces) are reached, are denoted by S_A and S_B; they subtend angles of $\pm(45 \pm \phi/2)$ with the *1*-direction. An important special case of Rankine states is shown in Fig.14. This figure represents the shear

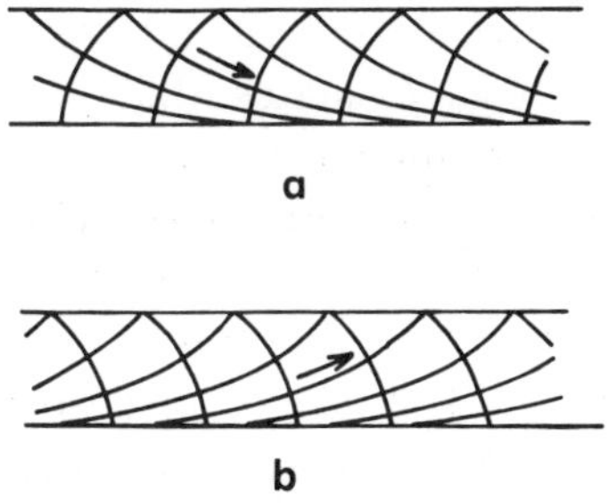

Fig. 14. Slip-line fields in a layer (c finite; $\phi = 0$) in active (a) and passive (b) Rankine flow. (After Nye, 1951.)

lines (i.e. traces of the shear surfaces) in a layer of material in which c is finite, but the angle of internal friction ϕ is zero (after Nye, 1951).

The above description has been proved valid for materials ranging from soil (Bjerrum, 1954) to stiff clays (Bjerrum and Lo, 1963) to highly plastic clays (Casagrande and Rivard, 1959). However, the values of the constants c and ϕ in the Coulomb equation vary greatly for different materials. In loose beds of sand and scree, one finds $c = 0$. In this case, ϕ is the slope angle beyond which a pile formed of such materials cannot be stable (angle of repose; Carrigy, 1970). For sand, one has $\phi = 33°$ (Hahn, 1969), for screes consisting of friable loess (Lohnes and Handy, 1968) and marls (Heim, 1932) it is $\phi \sim 25°$; of slate (Heim, 1932) 26°–29°; of limestone, 32°; of gneiss, 34°; and of syenite or granite, 35°–40°. In clays, the angle of internal friction is near zero. The cohesion c ranges from near zero in very wet clays, which may be practically liquid, to 10^8 dynes/cm^2 in air-dried clay (Scheffer and Schachtschabel, 1970).

The above discussion refers to materials that can be described by soil-mechanical methods. In *rocks*, failure shows a different pattern; failure in rocks is represented by a process of breaking which is called "fracture": discontinuities are suddenly formed which propagate rapidly through the material. Even at the present time the physics of fracture is a difficult and imperfectly understood subject. Much information on it has been collected in a seven-volume treatise edited by Liebowitz (1969 ff.). Regarding fracture of *rocks*, Rinehart (1966) has presented a useful review and Handin (1969) a historical discussion of the subject.

For practical purposes, a phenomenological fracture criterion due to Coulomb and Mohr is of most use. It states that two potential fracture planes exist in a triaxial compressive stress state. These fracture planes contain the intermediate principal stress direction and form similar angles $\beta \leqslant 45°$ with the maximum principal pressure. If tensional stresses are present, the fracture occurs normal to the largest principal tension. The above criterion seems to fit observations of rock failure in the laboratory fairly well (Handin, 1969). Neglecting the intermediate principal stress may not always be completely correct, but its effect is definitely not very great (Handin et al., 1967; Mogi, 1967). Just before failure, the phenomenon of volume dilatancy occurs (Reynolds, 1885): the bulk volume of the material increases owing to the appearance of minute hollow spaces.

It remains to quote a few threshold ("strength") values for the stresses at which rocks will yield by fracture. Benioff and Gutenberg (1951) have noted that, in *shear* failure, strength values are approximately 10^9 dynes/cm^2 for most rocks (marble, sandstone, limestone). The strength may also depend on the *rate* of stress application (Peng and Podnieks, 1972). For *tensile* failure the strength values of rocks are considerably lower and also depend on the relative directions of the maximum tension and the foliation (or bedding) planes (Youash, 1969). Therefore, parallel to the planes, the tensional strength may reach 10^7 dynes/cm^2, normal to them, perhaps one-half of this value. In *jointed* rocks, conditions may again be different. Methods for determining the strength *in situ* have been described by Müller (1963, p.269 ff.). Bieniawski et al. (1969) have reported on laboratory procedures for discovering the strength of jointed rocks. The *tensile* strength of jointed rock is naturally near zero for tension directions normal to the joint planes. Regarding the *shearing* strength of jointed rock, a theoretical discussion has been attempted by Patton (1966) and by Ladanyi and Archambault (1969, 1972). Accordingly, the clue to the behaviour of such rock is that there are several modes of failure; these take the forms of a shear plane, a shear zone and a kink band, respectively. The strength characteristics of each failure mode must be considered separately. Naturally, this type of analysis introduces many parameters whose values are chosen *a posteriori*. General statements about them cannot be made.

4.16. Progressive failure

When a material fails, discontinuities arise within it (by definition). Naturally, these discontinuities, in turn, affect the original stress state in the medium. Thus, if failure occurs, for instance, by the formation of a fracture surface, the tensile stresses across this surface must become zero. The newly created stress state may lead to stress concentrations which create further instabilities and fractures in the material. This phenomenon is called "progressive" failure (Terzaghi, 1962). The stress relocations may occur rapidly, leading to progressive "brittle" fracture; or slowly, leading to progressive "creep" failure (Haefeli, 1966, 1967).

The actual development of the stresses during progressive failure has been studied by Müller and Malina (1968) who devised a numerical finite element technique for following the phenomenon during its evolution. An application of progressive rock mass failure to the explanation of loess slumps has been made by Lutton (1971). Other applications of progressive failure will be discussed in this book as the need arises.

4.2. The stability of slopes

4.21. General remarks

Our main concern is with the *dynamics* of violent changes on the surface of the earth as these changes represent catastrophes. A related problem is the determination of the conditions under which such violent changes do *not* occur; i.e., the conditions of *stability* of the earth's topography.

Much more is known regarding the conditions of stability than about the actual course of a catastrophe. This is because in engineering practice (and most of our knowledge of the *quantitative* aspects of the features of the earth's surface come from practical engineers, not from pure scientists, the latter having been traditionally concerned with qualitative methods of descriptive natural history), stability is the aim; the occurrence of a catastrophe is only the consequence of some design error and its detailed investigation is of little interest to the engineer.

Seen from the standpoint of pure science, the course of a collapse is, of course, of interest. In this connection, the stability conditions are only the conditions which, if exceeded, will start the catastrophe. For this reason, however, they must be taken into consideration.

As in all the information on slopes in this chapter, a distinction has to be made between "soil"-mechanical and "rock"-mechanical conditions, as outlined in Sect.4.12. We shall discuss first the stability of "soil" slopes, then the stability of "rock" slopes.

4.22. Stability of soil slopes

Our aim is now to determine the stability of a slope bank consisting of "soil". As we have seen before, the strength of "soil" materials is governed by eq.4.15-1:

$$\tau = \sigma \tan \varphi + c \tag{4.22-1}$$

where τ is the shearing strength, σ the normal stress, c the cohesion and φ the angle of internal friction. We assume that the possible sliding surfaces are cylindrical so that the consideration of a single profile is sufficient and the problem can be considered as two-dimensional.

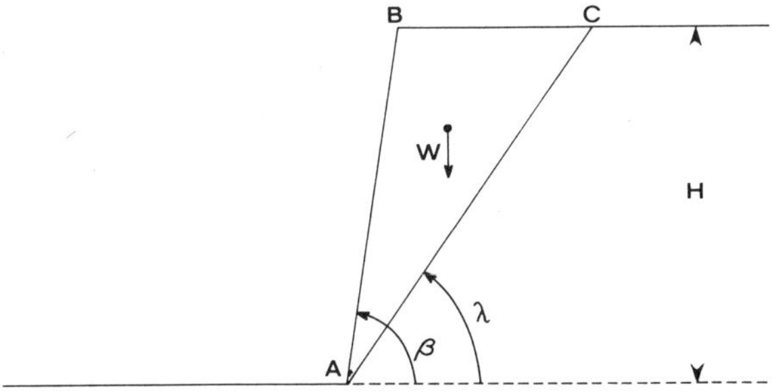

Fig. 15. Geometry of a slope bank with a plane sliding surface.

The simplest assumption for the shape of the potential sliding surface is that the latter is plane. We then have a geometrical layout as shown in Fig.15. In this figure, W denotes the weight of the slice ABC, H the height of the slope bank, β the slope angle and λ the angle between the sliding surface and the horizontal.

The slice ABC is just in equilibrium if:

$$W \sin \lambda = c \frac{H}{\sin \lambda} + W \cos \lambda \tan \varphi \tag{4.22-2}$$

The weight of the slice ABC (assuming unit thickness) is:

$$W = \rho g \frac{H^2 \sin(\beta - \lambda)}{2 \sin \beta \sin \lambda} \qquad (4.22\text{-}3)$$

where ρ is the density of the material and g the gravity acceleration. According to Fellenius (1927) the most dangerous position of the potential surface of sliding is that which requires the largest cohesion c for equilibrium to be maintained. Solving eq.4.22-2/3 for c and setting $\partial c/\partial \lambda = 0$ yields:

$$\lambda = \tfrac{1}{2}(\beta + \varphi) \qquad (4.22\text{-}4)$$

Inserting this into eq. 4.22-2 and solving for H yields the critical height h_c:

$$h_c = \frac{c}{\rho g} \frac{2 \sin \beta \cos \varphi}{\sin^2 \frac{1}{2}(\beta - \varphi)} \qquad (4.22\text{-}5)$$

The last formula is customarily represented by introducing a stability factor $N(\beta, \varphi)$:

$$h_c = \frac{c}{\rho g} N(\beta, \varphi) \qquad (4.22\text{-}6)$$

where:

$$N(\beta, \varphi) = \frac{2 \sin \beta \cos \varphi}{\sin^2 [\frac{1}{2}(\beta - \varphi)]} \qquad (4.22\text{-}7)$$

For $\varphi = 0$ (soil has only cohesion), N becomes particularly simple:

$$N(\beta) = 4 \cot(\beta/2) \qquad (4.22\text{-}8)$$

The critical height for a slope bank to remain just stable can be calculated from eq.4.22-6; the stability factor is represented in Fig.16.

The above analysis is only applicable to plane potential surfaces

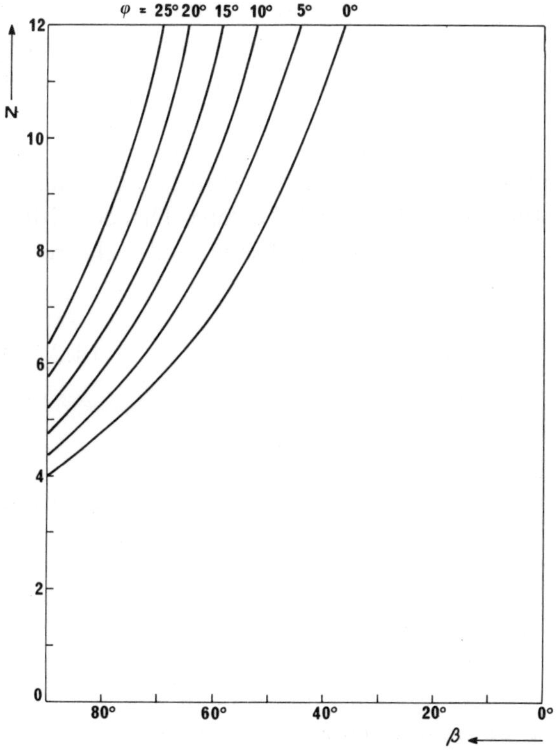

Fig. 16. Stability factor for various values of φ and β according to the model using a plane sliding surface.

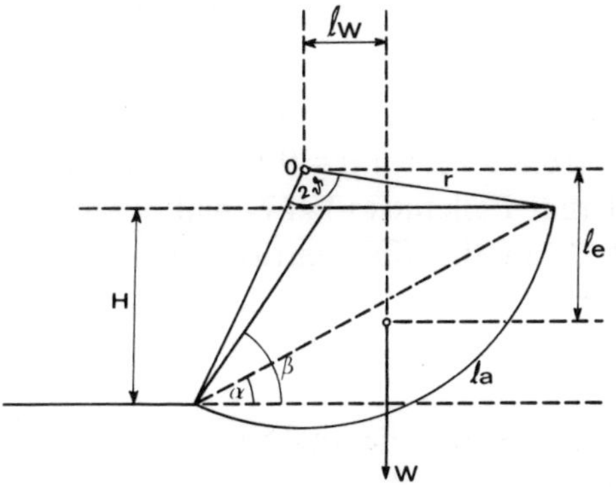

Fig. 17. Slope failure along a critical toe circle.

of sliding. A more sophisticated treatment is obtained if circular cylindrical sliding surfaces are used; the geometry is then as shown in Fig. 17. This method seems to have been proposed for the first time by Petterson (1916); more detailed calculations were made by Fellenius (1927, 1948) and by Taylor (1937, 1948).

The equilibrium condition requires that the sum of all moments around the centre of the circle O (Fig. 17) be zero. This condition can be easily expressed only if $\varphi = 0$. One has immediately, referring to Fig. 17:

$$Wl_{\mathrm{w}} - cl_{\mathrm{a}}r = 0 \tag{4.22-9}$$

or:

$$c = \frac{Wl_{\mathrm{w}}}{rl_{\mathrm{a}}} = \rho g H / f(\lambda, \alpha, \beta) \tag{4.22-10}$$

where f denotes "a function of". One obtains again the most dangerous surface of sliding as that for which the cohesion c required to maintain equilibrium is largest. The calculations can be made by a similar procedure to that employed in the case of a plane potential surface of sliding. Thus, one obtains the critical cohesion, and upon solving for H, the critical height h_{c} which can be represented by a new stability factor:

$$h_{\mathrm{c}} = \frac{c}{\rho g} N(\lambda, \alpha, \beta) \tag{4.22-11}$$

The stability factor was calculated by Fellenius (1948) and is shown in Fig. 18. It increases from 3.85 for $\beta = 90°$ to 7.35 for $\beta = 0°$. Here, it should be noted that Fellenius (1927) had obtained a value of $N = 8.37$ for $\beta = 0°$ in the first edition of his book. Although this earlier value was corrected by Fellenius (1948) in the later edition of his book, the incorrect value nevertheless was published in many textbooks.

The above analysis assumed $\varphi = 0$. The analysis for $\varphi \neq 0$ becomes very complicated. In order to take some account of the increase of the shearing strength τ with depth (actually due to the

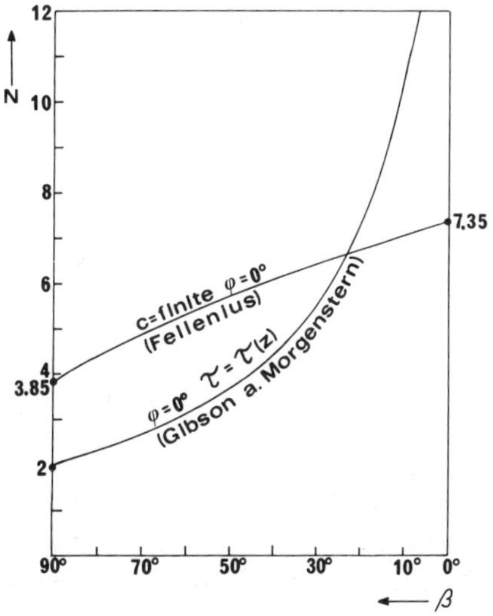

Fig. 18. Stability factor N for various cases.

overburden and a finite φ), Gibson and Morgenstern (1962) assumed a linear shearing-strength increase with depth:

$$\tau = kz/h \qquad (4.22\text{-}12)$$

where k is a proportionality constant, z the depth of the sliding circle below the top of the slope bank. For each supposed position of the potential sliding circle, a safety factor can be calculated as the ratio between the moment resisting displacement and the moment trying to effect displacement. The most dangerous surface of sliding can be found by a numerical trial-and-error procedure and a stability factor can be calculated by which the critical collapse height h_c is again represented by a formula of the type of eq.4.22-11. Assuming $\tau = c$ at the foot of the slope, the stability factor of Gibson and Morgenstern (1962) can be compared with that of Fellenius (1948) for $\varphi = 0$. The result is shown in Fig.18.

The above calculation is semi-empirical, because of the heuristic assumption of eq.4.22-12. For a true treatment of the case $\varphi = 0$, it is still hoped that the cohesion just required to maintain equilibrium can be expressed in the following form:

$$c = \rho g H \frac{1}{F(\alpha, \lambda, \beta, \varphi)} \tag{4.22-13}$$

Maximizing the cohesion and solving the equations for H would yield the critical height h_c. Again, one will attempt to express the latter as follows:

$$h_c = \frac{c}{\rho g} N(\alpha, \lambda, \beta, \varphi) \tag{4.22-14}$$

Nobody seems as of yet to have followed this principally possible scheme. Values of N calculated by Taylor (1937) and reproduced e.g. by Terzaghi (1943) were obtained in an entirely different manner (friction circles). These values, used considerably in practice, are shown in Fig. 19. Note, however, that the correct curve for $\varphi = 0$ has been given here.

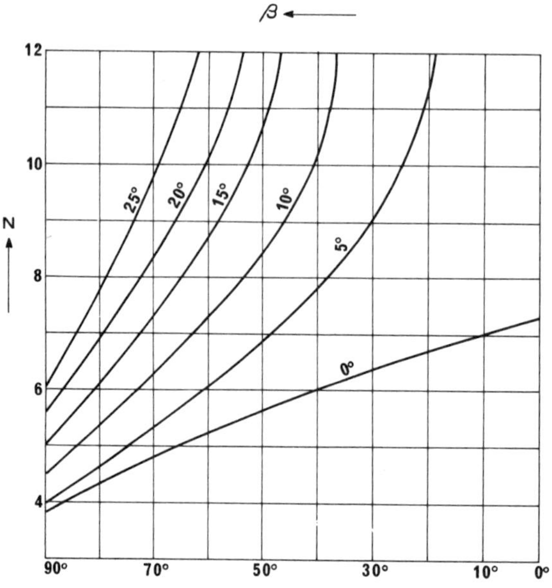

Fig. 19. Stability factors N as a function of slope angle β for various values of φ.

Fellenius (1948) has used another approximate method to obtain N by splitting the sliding body vertically into slices normal to the cross-section under consideration. The stress changes across slices were neglected. Later, Bishop (1965) introduced the correct stress condition between slices. Evaluations of such schemes (many slices) have to be effected on a computer.

The above theories assume that the potential surface of sliding is circular. Another shape which is frequently encountered in nature, is the chair-type shape. Often, a potential sliding surface of small inclination is indicated by the geological conditions. Then, a steep break occurs further up on the slope bank connecting the geological sliding surface to the day surface. Cases of this type have been described by Hutchinson (1969) in connection with landslides in Kent.

Even more general sliding surfaces have been investigated by Nonveiller (1965). Borowicka (1968) devised a simple and consistent procedure to calculate the stability for various shapes of potential sliding surfaces.

The general conditions for slope stability analysis have also been investigated on models (Schofield, 1971). For this purpose, centrifuges provide a possibility for varying the gravity acceleration.

Many case histories can be cited in which slopes have been designed by engineers according to the criteria mentioned above. For stability analyses, the influence of the pore water has to be taken into account, so that in the fundamental eq.4.22-1 the effective pressure (cf. Sect.6.32) has to be used and not the simple overburden pressure. A very great number of stability analyses have been published. Usually, the problem is to choose the correct quantities c and φ and to find the potential sliding surface. The piezometric heads in the slope bank may have to be known so that the effective pressure can be calculated, forcing one to solve first the problem of the flow of the groundwater in the soil. Of great interest are cases where it was found that, in order to obtain correct results, the cohesion c must be taken as nearly equal to zero and φ substantial (to 35°) for some clays (Mitchell and Eden, 1972). This is quite contrary to what is usually assumed for clays.

4.23. Stability of rock slopes

The equations of soil mechanics are no longer immediately applicable if the material of which the slope consists is "rock". As noted in Sect.4.12, the question of replacing "soil mechanics" with "rock mechanics" depends on the relative size of the phenomenon and the single rock elements in question. In addition, in stability calculations, the stresses are transmitted in the whole rock massif because of its strength which can lead to stress concentrations that are not observed in slopes consisting of material with a low shearing strength or a low cohesion. Finally, in conditions where "rock mechanics" must be applied, the global tectonic stress field can have an important influence.

In view of the above, the stability theory of soil mechanics applies not only when the individual rock inhomogeneities are small compared with the phenomenon in question, but also generally in low slope banks. Thus, Müller (1963, p.404) noted that for *low* slopes comprised of rocks up to a height of approximately 20 m the stability can be calculated essentially by soil-mechanical methods. The one difference, however, is that the potential sliding surfaces need no longer be critical toe cylinders, but surfaces of other geometrical shapes that may be predestined by the prevailing joints. Rock masses, in general, are highly anisotropic, particularly if they are layered. The prevailing stability conditions may cause the ubiquitously observed asymmetrical shapes of mountain chains with a steep "front" and a more gently inclined, often grassy "back". For the calculations of the shapes, slope angles and stable heights, only guesses and experience can give a clue. For the potential gliding surfaces that are considered, the mechanics of soil mechanics, based on eq.4.15-1 can be directly applied. The surface with the greatest likelihood of sliding will then be the critical surface for which the stability has to be determined. For this, one will have to use in situ values for the "effective" cohesion and angle of internal friction. A number of such values has been collected by Petzny (1967). On the joint surfaces, the cohesion may be near zero.

The above statements are affected if water is present in the rock joints. The procedure in this case is similar to that when water is present in unconsolidated materials: the water pressure is simply subtracted from the actual stresses; the remaining difference then represents "effective stresses". This means that the rock water is displacing the Mohr circles describing the stress states towards the zero-point, thereby increasing the danger of failure, just as this was the case in soil-mechanics considerations.

The above general considerations have been the basis of a number of stability investigations which will now be reviewed. These may be classified into: (i) purely empirical observations: (ii) two-dimensional stability calculations; and (iii) three-dimensional studies.

We begin with empirical investigations. These were particularly stressed by Bjerrum and Jørstad (1968) of the Norwegian Geotechnical Institute who analyzed some 300 unstable rock slopes. Qualitatively, it was found that the frost action, the chemical weathering, the temperature variations, the vegetation, the direct action of water and the influence of the snow cover make it impossible for steep jointed rock slopes to exist stably for any length of time, unless they are covered by a protecting layer of scree or have a slope angle which is not greater than the angle of repose of the corresponding loosely packed debris material (about 40°). Thus, according to Bjerrum and Jørstad (1968), the stability factors of rock slopes cannot be determined by theoretical calculations because the relative importance of the factors mentioned above cannot be estimated. According to these authors, the best procedure is, therefore, to make comparisons between unknown and known slopes.

Nevertheless, the empirical approach does not provide any insight into the mechanics of the stability (or instability) of slopes. Therefore, theoretical studies are desirable, even if they have not yet been able to yield any reliable predictions. As noted, such studies may be based on a two-dimensional or three-dimensional geometry.

A simple two-dimensional analysis can be made by assuming that there are pre-existing joints in rock walls. If a single joint intersects a slope bank the geometry shown in Fig.20 is obtained. The condition for the block sliding above the joint is:

$$W \sin \psi = W \cos \psi \tan \delta \qquad (4.23\text{-}1)$$

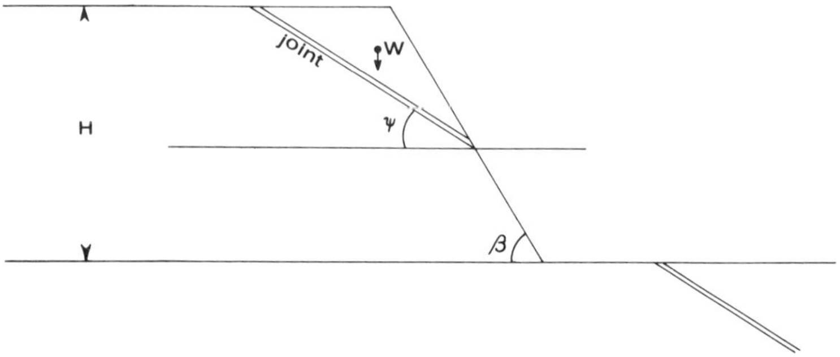

Fig. 20. A joint intersecting a slope bank.

where W is the weight of the block and δ the angle of friction at the joint. A somewhat similar geometry was studied by Wittke (1964), who assumed the tensional strength to be zero across the joints which permitted the slope to be of a more complicated form.

For high (above 80 m) and steep banks the weight of the rock masses causes stress concentrations at the foot of the slope. It has been noted for some time that the so-called weathering decay of rock walls starts at the foot of the wall, not at the top, due to stress concentrations. Gerber and Scheidegger (1965, 1969) have shown numerous examples of this. The stress concentrations may only be caused in part by the weight alone, but also by the action of the ever-present tectonic stress field. For high slopes, Terzaghi (1962) has given an elementary formula which gives the critical height H as a function of the stability factor N for the geometry shown in Fig.21:

$$H = \frac{c}{\rho g} N \tag{4.23-2}$$

with:

$$N = \frac{1}{\cos\epsilon\,(\sin\epsilon - \cos\epsilon\tan\delta)} \tag{4.23-3}$$

where ρ is the density of the rocks, c the cohesion and δ the angle of friction across the joints, and ϵ the angle of inclination of the

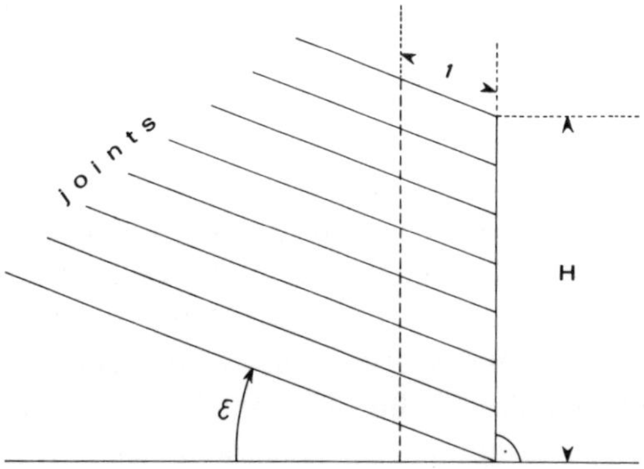

Fig. 21. Large, high rock wall with joints.

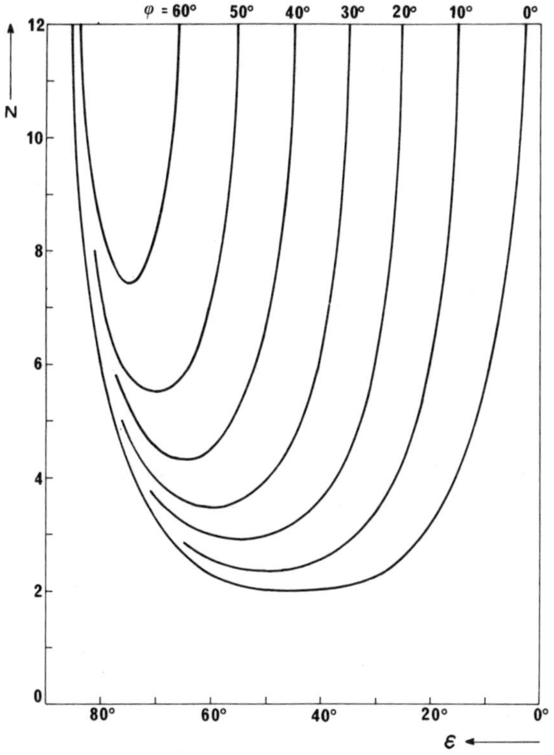

Fig. 22. Stability factor for high rock walls.

joint. If N is graphically represented, one obtains the result shown in Fig.22.

Finally, there are the attempts to calculate the equilibrium of a rock mass in three dimensions, as it is bounded by the slope surface, the joints and the bedding. This approach was taken by Londe (1968) and his co-workers (Londe et al., 1970). The forces on such a rock mass, tetrahedral in shape, are all taken into account and the stability for the mass is then calculated. Earlier, Wittke (1964) used a simpler geometry (two planes intersecting a slope bank, so that a wedge is formed), but his assumptions were similar to those of Londe. Accordingly, the tensional strength is assumed as zero across the joints and the bedding planes (= "tectonic" planes), so that only friction can act there. The effective angles of friction on these planes are empirical constants. It is then possible to calculate the equilibrium conditions for a block between bedding planes; these conditions depend only on the direction in space of the resultant of the weight (and possibly other loads) and of the hydrostatic forces due to water in the "tectonic" planes. If there is to be stability, this resultant must be oriented so that either no motion is kinematically possible or that motion can be balanced by the frictional forces (i.e., it must lie in the corresponding friction cones if sliding is kinematically possible). This condition yields certain domains of directions of the resultant in space, best represented on a unit sphere (or on its plane projection), for which there is a state of equilibrium and for which there is also a state where no equilibrium exists. The method thus obtained yields a means of estimating the stability of single "elemental" masses of rock on a slope as they are bounded by the ever-present joint system and bedding planes. The geometry of the latter has to be found through direct observations in the field.

4.24. Quasistatic adjustment of slopes

When slow external alterations occur in the geometry of a slope, usually through human activity or river erosion at the slope's foot, the stability conditions may eventually be violated and the slope will collapse (cf. Fig.23). This may be a sudden, rapid event, such as a land slide. The latter phenomenon will be studied in detail later.

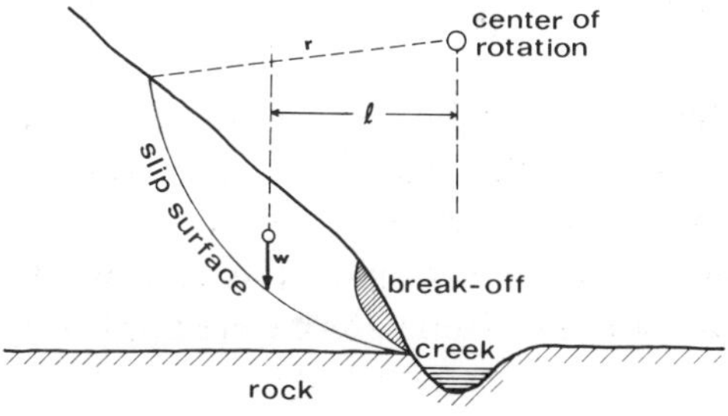

Fig. 23. Erosion at the foot of a slope by a creek can cause instability.

However, the adjustment to the geometrical changes can also be continuous or quasicontinuous. Certainly, violent adjustments may occur continually at intervals, as the stability conditions are exceeded, but viewed on a temporally large scale, the phenomenon appears as "quasistatic" adjustment to the changing conditions.

Thus, if the foot of a slope is eroded, the latter, in soft material, may simply be expected to recede, maintaining in the mean the declivity which corresponds to the angle of repose of the material. However, progressive failure may be initiated (Lo and Lee, 1973).

In rock, the slope recession is governed by the cleavage system. The resulting patterns, therefore, depend entirely on the type of jointing that is present. It is very difficult to predict these patterns theoretically, but model experiments have been made by Hofmann (1972) and by Fürlinger (1972) for various types of cleavage patterns.

Hofmann (1972) used a frame in which he represented a slope bank by square blocks. The erosion was modeled by the removal of blocks one by one at the foot of the slope. It was interesting to observe that tension fissures appeared far back in the slope bank; the front showed a downward rotation.

Fürlinger (1972) modeled the cleavage pattern that is characteristic for thinly layered media, using IBM cards as units. A characteristic downward bending of the layering towards the slope bank was observed.

The continuous erosion on a slope also can cause stress build-up within it, leading to exfoliation (Müller, 1970). This will be treated separately.

4.25. Slopes and earthquakes

When investigating the effect of an earthquake on slope stability the prime effect noted is caused by the ground acceleration. According to D'Alembert's principle of inertial forces, the effect of an acceleration $\ddot{x}$ of the frame of reference upon the motion of a mass m is the same as if a force of magnitude $-m\ddot{x}$ were applied to that mass.

Hence, in all stability considerations, one simply has to add a force $-m\ddot{x}$ to the forces acting on the mass m where stability is being investigated; $\ddot{x}$ is the acceleration of the ground due to the earthquake. Referring to Fig.17, one can define a "safety factor" G for a slope bank as follows (quotient of resisting moment and driving moment):

$$G = \frac{c l_\mathrm{a} r}{W l_\mathrm{w}} \tag{4.25-1}$$

For $G < 1$, a slide will occur. Consequently, if an earthquake produces a horizontal acceleration $\ddot{x}$, the safety factor G' becomes:

$$G' = \frac{c l_\mathrm{a} r}{W(l_\mathrm{w} + l_\mathrm{e}\ddot{x}/g)} \tag{4.25-2}$$

The ground accelerations $\ddot{x}$ found in earthquakes have been estimated in Sect.2.33; maxima lie around 1.25 g. Thus, an estimate of the stability of a slope in a particular earthquake can be made. In the above equation, the effect of the pore pressure must be added, as there may be a significant effect upon the stability owing to seepage of ground water (Newmark, 1965).

The analysis based on D'Alembert's principle is essentially a static one. It was pointed out by Finn and Miller (1973) that a dynamic analysis would yield more accurate results, inasmuch as the short duration and the change of direction of the shaking force may elicit an entirely different response from the ground than a static force. Such a dynamic analysis can be made by finite element techniques.

Furthermore, liquefaction may occur owing to an earthquake;

i.e., not only may the effect of the ground motion make itself felt through eq.4.25-2, but the cohesion *c* may be lost. This may occur in saturated sandy soils (Bazant, 1966; Seed, 1968; Kishida, 1969; Ishihara and Yasuda, 1973). In such materials, the contact between the grains is lost by the shaking which will lead to an appearance of thixotropy in the substance. Qualitatively, the physical processes occurring during soil liquefaction are quite well understood (Seed and Lee, 1966; Kurzeme, 1971): the saturated soil tends to compact under the earthquake shaking which leads to an increase in pore water pressure; this reduces the shearing resistance in conformity with eq.6.32-2. A theoretical analysis of the process was made by Shibata and Yukimoto (1970) whose final equation, however, still contains undetermined parameters. Experiments to determine the conditions under which soils become liquefied by cyclic loading have been performed by Peacock and Seed (1968), Ambraseys and Sarma (1969), Finn et al. (1970), Silver and Seed (1970), Tanimoto and Suga (1971), and Seed and Peacock (1971). The results of these experiments are difficult to state in general terms; plots have been obtained for specific soils, e.g., of the number of cycles required to achieve liquefaction in dependence of the cyclic shearing stress and of the original confining pressure (Peacock and Seed, 1968). However, Finn et al. (1970) noted that the liquefaction process does not only depend on the prevailing conditions, but also on the previous strain history. Ambraseys and Sarma (1969) found that the liquefaction conditions are essentially determined by a critical velocity. A mass of waterlogged sand, therefore, can lose its stability under relatively small shakings (Bjerrum et al., 1961).

Cases of landslides triggered by earthquakes have been reported on many instances. Particularly notable reports were made by Morton (1971) referring to California and by Kobayashi (1971) referring to Japan.

4.3. Pure rock phenomena

4.31. Description

Accidents caused by rocks are usually due to the sudden failing of the rocks. Depending on the rock mass, greater or lesser catastrophes may be the result.

Whether a rock mass fails or not depends upon its stability. Therefore, the general methods of stability calculations outlined in Sect.4.23 are of paramount importance. These may, however, be somewhat affected by the presence of "internal" stresses.

A rock fall, once a rock mass has failed, is a simple gravitational phenomenon. It may engender a rock slide on a slope below the break-off point.

The internal stresses in rocks may cause large sheets to split from a rock surface; this phenomenon is called exfoliation. If it occurs with explosive force it is termed a "rock burst". Rock bursts mostly occur in underground openings and can cause much damage and many deaths.

4.32. Tectonic stresses and rock masses

Naively, one would assume that the stresses in rock masses are compressive and that this compression is caused by the overburden. On the average, the overburden pressure p is:

$$p = 0.23\ \mathrm{kp/cm^2}\ \text{per m of depth} \qquad (4.32\text{-}1)$$

Unfortunately, this "theory" is slightly too simple. It has been indicated that stresses in rocks are non-isotropic (i.e., that the 3 principal stresses are unequal). This can be inferred from the tectonic features such as faults and folds seen in many areas and also from direct measurements.

Even if the existence of "tectonic" stresses is discounted, the overburden would produce lateral stresses because the material, when buried, cannot expand laterally. The lateral strain ϵ_x in a stress field with principal stresses σ_x, σ_y and σ_z (x, y horizontal, z vertical Cartesian coordinates) is:

$$\epsilon_x = \frac{1}{E}(\sigma_x - m(\sigma_y + \sigma_z)) \qquad (4.32\text{-}2)$$

where E is Young's modulus and m Poisson's ratio. Since the lateral strain must be zero, one obtains:

$$\sigma_x - m(\sigma_y + \sigma_z) = 0$$

With $\sigma_y = \sigma_x$, this yields:

$$\sigma_x = \sigma_y = \frac{m}{1-m}\sigma_z = \frac{m}{1-m}\rho g z \qquad (4.32\text{-}3)$$

The lateral "pressure" should be $m/(1-m)$ times the overburden pressure; this is less than the latter: the fact that the material is solid and therefore heavy explains this. For $m = 0.25$ (a commonly found value) one has:

$$\sigma_x = \sigma_y = \tfrac{1}{3}\sigma_z \qquad (4.32\text{-}4)$$

As noted, the above calculation discounts the existence of tectonic stresses. The tectonic stresses are superposed on the overburden effect. The basic idea is that tectonic motions take place slowly. Thus, there is creep, but during the creep, specific stress states are continuously regenerated. However, in each instant the system is in elastic quasi-equilibrium. Near the surface there is a boundary condition: the surface cannot transmit any shear. Hence, the principal stress directions near the earth's surface have to be vertical or horizontal. Depending on the orientation of the maximum and minimum principal stresses, three standard stress states are possible (Anderson, 1942; see also Scheidegger, 1963a). The tectonic stress state is usually uniform over large areas. It is important to know its character in an area where the stability of rock masses has to be assessed, inasmuch as the intrinsic stresses in rocks may be responsible for sudden failures.

An important effect of the tectonic stresses is that all rock masses have become fractured to some degree: conjugate "systems" of joints have been created which represent the Mohr-fracture surfaces corresponding to the tectonic stress state which prevailed at the time of their creation. The joints found in rock masses are of fundamental importance in stability calculations (cf. Sect.4.23). The stress state responsible for the creation of a system of joints can often be reconstructed from their analysis.

If the overburden of a buried rock mass is removed, internal lateral stresses corresponding to the buried state (cf. eq.4.32-3) will often remain. In effect, most rocks have been in a different stress state from that in which they are seen now. Because the earlier stress does not completely disappear through relaxation, high internal stresses are generally present. Varnes and Lee (1972) have described an analog model exhibiting the mobilization of such residual stresses when a new free surface is created in a rock mass that appears to be in equilibrium. This type of effect may be the cause of many sudden accidents.

4.33. Rock falls

Rock falls occur commonly on steep rocky walls. A fall itself may consist of single rocks, or of a whole mass: tens of thousands of cubic metres may suddenly break from a wall and thunder into the space below.

Whilst the fall in itself represents an event of considerable force, the events following it are also of utmost importance. Usually, a steep rocky wall is bounded at the bottom by an inclined apron of screes which has been built up by the continuous addition of material from the wall above. A rock fall may continue along this inclined apron some distance as a *debris avalanche*, or single rocks may detach themselves, begin rolling or jumping and may continue downhill for a long distance, even beyond the confines of the apron of scree. The mechanics of these various phenomena are entirely different from each other. Here, only the fall itself will be considered; the mechanics of the subsequent phenomena will be discussed in Sect.4.4.

Whether and how a rock fall occurs on a wall is a problem of static stability. Since one is dealing with *rock* slopes, the methods of rock mechanics have to be applied for a stability analysis. This is different from the usual slope stability analysis commonly practised in soil mechanics, where the potential sliding surface is assumed to be a critical toe circle. In rock, the potential sliding surfaces are predestined by the cleavage system which is always present therein. This leads to potential break-out niches (cf. Sect.4.23). A stability analysis must be performed by calculating the equilibrium conditions for the possible break-out niches; the form of the niches must be based on a survey of the cleavage patterns present on the particular rock slope under consideration (cf. Müller, 1963).

When a rock falls, it gathers speed. The velocity on impact below is determined by the well-known laws of free fall. Thus, if H is the distance of the fall, the velocity v can be calculated:

$$v = (2gH)^{1/2} \tag{4.33-1}$$

where g is the acceleration of gravity.

Protective measures against a threatening rock fall must primarily be directed towards a stabilization of the rock mass; the application of rock bolts is the method generally used. The calculation of the number of rock bolts required must be based on an analysis of the size of the rock mass, as delineated by the joint system, that is likely to become unstable, and on an analysis of the prevailing internal rock stresses and water conditions. Much, however, is left to experience in this connection. Netting to catch falling rock has also been employed. Ritchie (1966) studied a number of measures to protect highways from falling rocks and reported that the most effective methods of protection were trenches and check fences. These must be designed so as to be effective against the expected rock masses and their trajectories.

4.34. Exfoliation

Exfoliation is a widespread phenomenon in nature. It encompasses features ranging from sheeting on the domed granite mountains of Australia, America and Africa, to horizontal fissuring in bedrock and to the splitting off of plates in sandstone (Matthes, 1930; Kieslinger, 1958; Jahns, 1943; Twidale, 1964).

Quantitative investigations of exfoliation show that the phenomenon is usually of pre-Glacial age and that it is independent of the primary structure of the rocks. The thickness of the plates increases with depth below the surface (cf. Fig.24) and the ordered orientation of the exfoliation joints disappears completely at a depth of about 50 m (in granite).

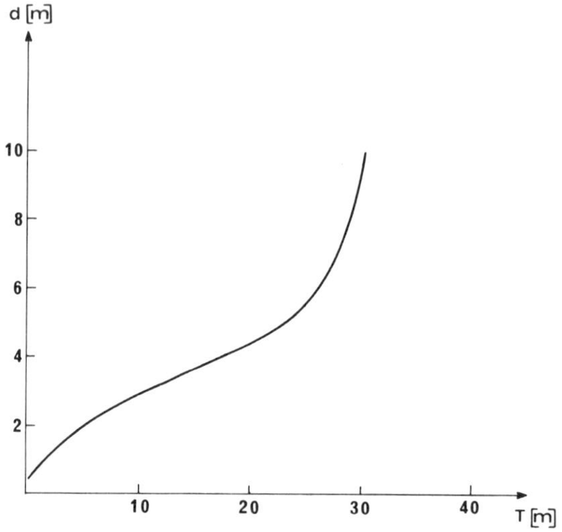

Fig. 24. Dependence of the plate thickness d on the minimal depth T (m) from which this thickness may be observed. (After Jahns, 1943.)

A critical review of the literature (Brunner and Scheidegger, 1973) shows that the most common explanation of exfoliation is a stress relief theory, according to which a removal of overburden would lead to exfoliation. However, if a simple model of this process is considered (Brunner and Scheidegger, 1973), it is immediately seen that it is difficult to obtain the required instability: if one calculates the potential buckling of an elastic plate under horizontal compression (caused by the former overburden), then the critical dimensions of the plate do not become reasonable.

However, it is well known that large horizontal stresses may exist in many places in rocks. According to Voight (1969), these may attain 100 kp/cm^2. If the exfoliation is considered as an analogon to the well-known multiple axial tension fractures observed in compression tests (the tensional stresses are induced stresses), a suitable model is found for it. The tensional stresses are induced in a uniaxial compression at the boundaries of the (Griffith) cracks; they can attain similar absolute values as the compressive stress. As soon as the induced tensional stresses exceed the tensional strength of the rock at the most dangerous points of the cracks, a progressive tension fracture occurs which is aided by notch effects. This initiates the exfoliation. The occurrence of the fracture pattern can be prevented by a little lateral pressure.

Let us assume that the cracks in the rocks have the form of ellipses whose axes are oriented parallel to the principal stress directions. The tension tangential to the ellipse is then given by (Isaacson, 1962):

$$\sigma_t = \sigma_1 + \sigma_3 - \frac{[(a-b)(\sigma_1+\sigma_3)-(a+b)(\sigma_1-\sigma_3)](a\sin^2\vartheta - b\cos^2\vartheta)}{a^2\sin^2\vartheta + b^2\cos^2\vartheta} \tag{4.34-1}$$

where the axis parallel to σ_1 is a; the angle ϑ is counted from the σ_1 direction.

Assuming σ_1 as a given horizontal stress, σ_3 as the overburden pressure:

$$\sigma_3 = \rho g D \tag{4.34-2}$$

where D is the thickness of the overburden, one finds for the value D_0 at which the induced tension σ_t is equal to σ_1:

$$S_T + \sigma_1 = \rho g D_0 \left(1 + 2\frac{a}{b}\right) \tag{4.34-3}$$

where S_T is the tensile strength of the rock. Hence:

$$D_0 = \frac{\sigma_1 + S_T}{\rho g \left(1 + 2\frac{a}{b}\right)} \tag{4.34-4}$$

Fig.25 represents curves for D_0 as a function of $(\sigma_1 + S_T)$ for $\rho_2 = 2.5\ \mathrm{g/cm^3}$. Taking $\sigma_1 = 100\ \mathrm{kp/cm^2}$ and setting $S_T = 60\ \mathrm{kp/cm^2}$, one finds $D_0 = 50$ m. At this depth the induced tension equals the overburden pressure, so no exfoliation can be initiated beyond it. The model explains the phenomenologically observed features of exfoliation.

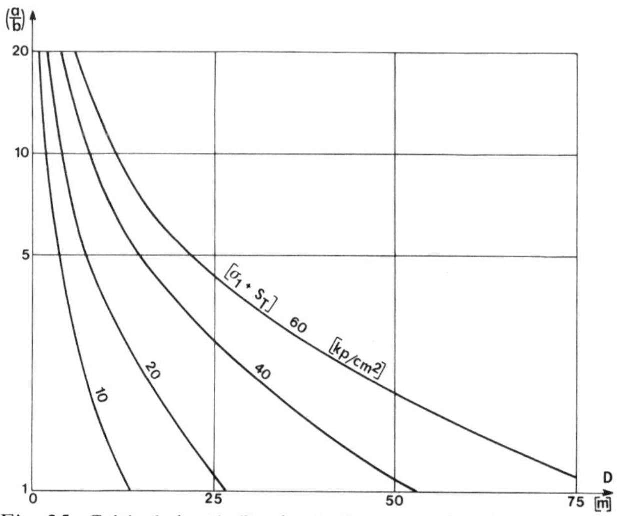

Fig. 25. Critical depth D_0 for induced tension fractures in dependence of the ratio of the axes of the ellipse and on $(\sigma_1 + S_T)$. (After Brunner and Scheidegger, 1973.)

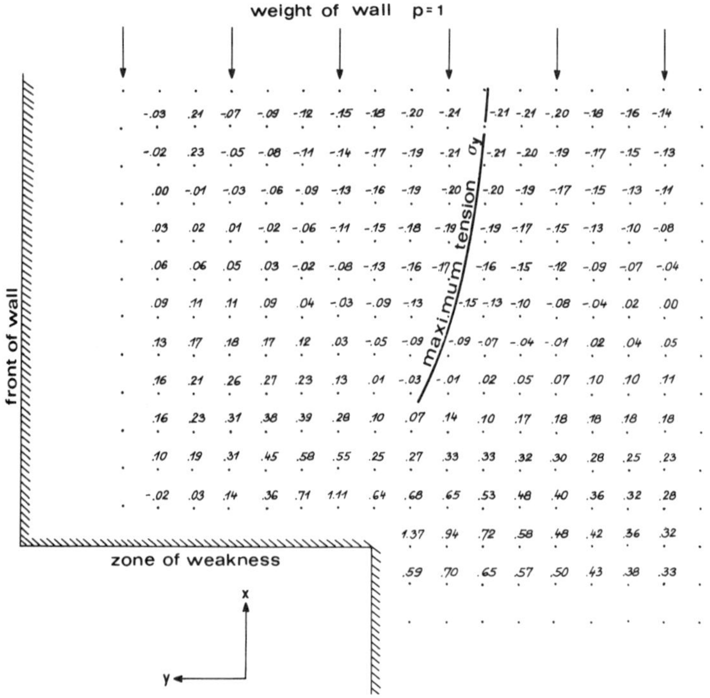

Fig. 26. Line of maximum tensional stress normal to the surface of a wall. (After Brunner and Scheidegger, 1973.)

For an explanation of the exfoliation of plates parallel to a vertical (or very steep) rock wall, a different model must be invoked. At the weak zone, which is always induced by stress concentrations (Sturgul and Scheidegger, 1967) at the foot of the wall a tension fracture starts and progresses behind the wall parallel to its surface as shown in Fig.26.

4.35. Rock bursts

A rock burst is caused by the failure of a surficial rock mass and occurs with explosive violence. It occurs almost invariably by shear failure and may extend from the "spitting" of a rock face to the collapse of a whole area. It occurs mainly on new man-made rock faces, particularly in underground mines. In this instance, it is only conditionally a "natural" catastrophe; it occurs mostly as a "natural" (unplanned) follow-up to man-made changes in the environment.

The mechanism of rock bursts seems to be based on a sudden release of accumulated strain energy. In a volume V subject to the principal stresses $\sigma_1, \sigma_2, \sigma_3$ the strain energy ϵ/V is:

$$\epsilon/V = \frac{1}{E}(\sigma_1{}^2 + \sigma_2{}^2 + \sigma_3{}^2 - 2m[\sigma_1\sigma_2 + \sigma_2\sigma_3 + \sigma_1\sigma_3])$$

where E is Young's modulus and m Poisson's ratio. Tremendous energies are present in each rock volume already at shallow burial-depth. In an unsupported underground span, the strain energy varies with the 5th power of the width of that span (Isaacson, 1962, p.160).

A rock burst, then, is caused by high stress concentrations. Often, a rock is in a highly unstable equilibrium, and some small trigger effect will set off a major rock burst. Such trigger causes would be, for example, routine blasting in mine workings, temperature changes, and changes in moisture (through its effect on rock stability).

A prediction of rock bursts can be made only upon a probabilistic basis; in this respect, bursts are very similar to earthquakes. An increased microseismic activity may herald an impending instability. The rate of creep in rock is normally logarithmic; a deviation from normal logarithmic movement rates may be a sign of an imminent burst.

As preventive measures there exist, of course, the general stabilizing techniques involving rock bolts. If a face has to be worked, destressing techniques have been successful: preliminary blasting is made in the highly stressed area behind the face so there will be less stress on the face itself.

4.4. Accidents on scree and talus accumulations

4.41. Description and classification

Scree and talus accumulations are commonly found in mountainous areas. They evolve as debris covers over rock below steep walls or in a cone shape formed by the accumulation of material from gulleys. The phenomena considered in this section occur without water having any significant effect.

A scree slope is a feature which is in dynamical equilibrium and it slowly evolves by the addition of material from above. The slope angle corresponds to the angle of repose of the material. The mechanical processes that may lead to accidents on such slopes are the following.

(a) Internal creep. The material of the slope is in a limiting Rankine state; hence slow motion along the slip lines must be assumed to occur. While such motion does not represent a catastrophic event, it may, nevertheless, lead to an instability that can trigger a collapse.

(b) Debris slides. A scree slope responds to a disturbance of the dynamical equilibrium by a local slippage of material which may then fall some distance downhill as a debris slide. Such slides can also be triggered by rock falls from above. They can cause considerable damage and loss of life.

(c) Rolling and jumping stones. Single stones may detach themselves from a debris slide and fall a long distance downhill. Such stones may act like bombs.

4.42. Scree and talus creep – stability

The internal creep in a scree slope is intimately bound with the stability of such a slope. The general theory of stability of slope banks has already been given in Sect.4.15; we shall present here only the applications of the stability theory to scree slopes.

For loose materials, the cohesion c is zero in eq.4.15-1. The slope angle β of a scree slope is close to the angle of repose, which is equal to the angle of internal friction φ. Values for various materials have been given in Sect.4.15. In addition, there may be a dependence on particle size (Behre, 1933) as shown in Table VI. Chandler (1973) claims that in *really* dry slopes the angle φ is at least 39° and that the lower values commonly found heuristically are caused by the presence of water.

TABLE VI

Dependence of the angle of repose on the particle diameter

Diameter	Angle of repose
30 cm	35°
15 cm	35°
10 cm	32° – 34°
5 cm	31° –32°
1.2–2.5 cm	26° –31°

When material accumulates on a slope, it may become unstable. In order to calculate the stability of a slope, the theory of Fellenius (Sect.4.15) is generally used. However, simple analytical expressions can only be found if $c \neq 0$, $\varphi = 0$. This case does not correspond to that of slopes. However, the cohesion can be replaced by the average shearing limit prevailing in a layer of scree (Gerber and Scheidegger, 1974); one has at depth x:

$$\tau(x) = \sigma_n(x) \tan \varphi \qquad (4.42\text{-}1)$$

where σ_n is the normal pressure.

However, σ_n is caused by the weight W of the overburden:

$$\sigma_n = W \cos \beta = x \rho g \qquad (4.42\text{-}2)$$

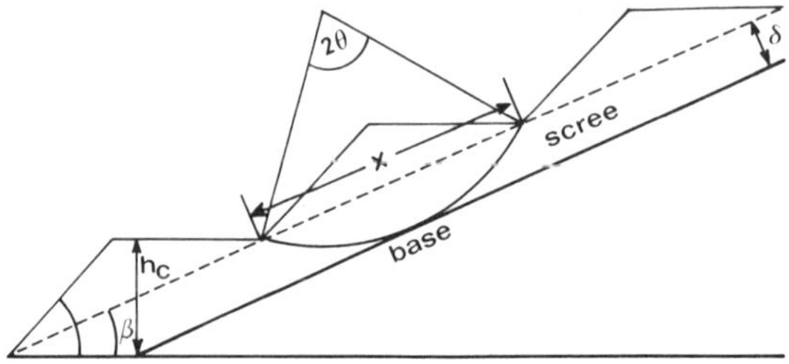

Fig. 27. Definition sketch for the calculation of the stability of a scree slope.

whence:

$$\tau(x) = \rho g x \tan \varphi \tag{4.42-3}$$

The average shearing limit in a layer of thickness δ is thus:

$$\bar{\tau} = \frac{1}{2} \delta \rho g \tan \varphi \tag{4.42-4}$$

which is substituted for the cohesion in calculations of scree-slope stability. If one assumes that slope instability is caused by a step-shaped accumulation of material (cf. Fig.27), one obtains from the general slope-stability eq.4.22-14:

$$h_c = \frac{1}{2} \delta \tan \varphi N \tag{4.42-5}$$

This allows for a stable section X of a slope:

$$X = \frac{\delta}{2} \frac{\tan \varphi}{\sin \beta} N \tag{4.42-6}$$

For $\beta \sim \varphi \sim 30°$ one obtains for X about 3.7 δ which corresponds to the order of magnitude that is observed in nature.

Sections of length X will collapse as miniature "landslips" on a slope, thereby forming tongue-shaped overlapping deposits. In addition, an internal motion may take place along the slip lines, because temperature effects create an alternate contraction and expansion of the rock (Scheidegger, 1961) or because the continuing diminution of the debris pieces changes the angle of repose (Wood, 1942).

4.43. Debris slides

When part of a scree slope becomes unstable or a rock fall occurs from above, a debris slide may be triggered. Such slides proceed rapidly downhill, eventually come to a stop and form tongue-shaped deposits.

The rapid motion of a small mass down a scree slope is governed by the laws of dry friction. The coefficient f of such friction is close to the tangent of the angle of repose of the sliding debris mass which, in turn, is equal to the angle of internal friction φ (in the Coulomb-Mohr sense) of the material in question. Therefore, as can be shown easily, the downhill motion of a debris mass is independent of its size, because mass is eliminated in the equations of motion.

In a model based on friction, the total vertical height of the path of the slide divided by the total horizontal reach represents nothing but the average coefficient of friction f (Scheidegger, 1973) provided the slide started from rest. The deduction of this fact is evident from an inspection of Fig.28. Accordingly, if a mass m slides through the distance Δs, the energy principle requires:

$$\Delta(\tfrac{1}{2}mv^2) = mg\ \Delta s \sin \beta - \Delta s\ mgf \cos \beta \qquad (4.43\text{-}1)$$

where v is the sliding velocity and the meaning of the other sym-

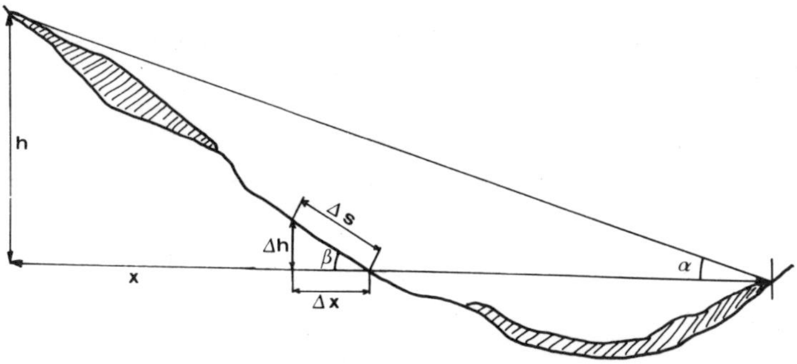

Fig. 28. Geometry of a slide. (After Scheidegger, 1973.)

bols becomes evident from Fig. 28. Now, the following relations hold:

$$\Delta s \sin\beta = \Delta h \quad (4.43\text{-}2)$$

$$\Delta s \cos\beta = \Delta x \quad (4.43\text{-}3)$$

whence:

$$\frac{1}{g}\Delta(\tfrac{1}{2}v^2) = \Delta h - f\Delta x \quad (4.43\text{-}4)$$

If we integrate the last equation from $s = 0$ to s = total course of the slide, assuming that the slide started from rest and is at rest at the end of its course, we obtain:

$$0 = h - fx \quad (4.43\text{-}5)$$

or:

$$f = h/x = \tan\alpha \quad (4.43\text{-}6)$$

The above calculations have to be modified if the slide does not start from rest but is, e.g., triggered by a rock fall. Then an initial velocity v_0 has to be introduced into the equations. The pertinent condition is:

$$\tfrac{1}{2}mv_0^2 + mgs\sin\beta - mgs\,f\cos\beta = 0 \quad (4.43\text{-}7)$$

where s is the downhill distance after which the slide comes to a stop. Solving for s, one obtains:

$$s = \frac{\tfrac{1}{2}v_0^2}{g(f\cos\beta - \sin\beta)} \quad (4.43\text{-}8)$$

If one assumes that the initial velocity v_0 is due to the energy of rocks falling from a wall of height H above the top of the scree slope, one has:

$$v_0^2 = 2gHK \quad (4.43\text{-}9)$$

where K is the part of the energy which is not lost upon impact ($K \leqslant 1$). Thus, one obtains the maximum distance s through which the debris slide can travel:

$$s = HK(f \cos \beta - \sin \beta)^{-1} \qquad (4.43\text{-}10)$$

The above equation shows again that the rock mass involved in the rock slide is immaterial. More energetic rock slides will travel farther than the less energetic ones. The slope angle β will be exactly that angle at which the energy loss by friction is compensated by the energy gain during the downhill motion.

It remains to estimate the quantities f and K occurring in the equations given above. As noted, the coefficient of friction is close to the tangent of the angle of repose φ of the material in question. One will find values from 0.6 to 0.9 for f. The quantity K seems to be approximately 0.25. With these values, the course of a debris slide can be calculated. The theory given here seems to be valid for debris slides whose volume does not exceed about 100,000 m^3. For larger slides, the coefficient of friction is found to be much smaller than the values given above. The problem of large, catastrophic landslides will be discussed in Sect.4.74.

4.44. Rolling stones

Whilst the effect of a debris slide triggered by a rock fall is confined to the scree slope itself, this is not true for the effect of individual blocks.

In fact, single blocks can travel very far. There is no absolute protection from such blocks: in the extreme case a block would transform the entire energy from its fall over the vertical distance H elastically into "projectile" energy. The velocity of such a projectile is given by:

$$v_0 = (2gH)^{1/2} \qquad (4.44\text{-}1)$$

and the maximum distance between impacts on a horizontal plane would be $s = 2\,H$.

As stated, the above example represents an extreme case. Experience indicates that, once loose from a slope, rocks soon start rolling; they do not jump appreciably. Thereby they lose energy.

It is possible to study the process of rolling stones by using a model of spheres rolling down a plane. Let us assume that the linear velocity of a rolling body be v. The kinetic energy E is then:

$$E = \frac{1}{2}mv^2 + \frac{1}{2}\Theta\omega^2 \tag{4.44-2}$$

where m is the mass, Θ the moment of inertia and ω the angular velocity of the rolling sphere. We note that:

$$\Theta = mk^2 \tag{4.44-3}$$

where k is the radius of inertia. For a sphere of radius r:

$$k^2 = \frac{2}{5}r^2 \tag{4.44-4}$$

Furthermore:

$$\omega = v/r \tag{4.44-5}$$

Thus (for a sphere):

$$E = \frac{1}{2}mv^2\left[1 + \frac{k^2}{r^2}\right] = \frac{1}{2}mv^2 \cdot \frac{7}{5} \tag{4.44-6}$$

Assume now that a rock travels with linear velocity v_A at point A on a slope, and with velocity v_B at point B. The points have the distance Δs along the slope (slope angle β) from each other and the difference in altitude ΔH. Evidently:

$$\Delta H/\Delta s = -\sin\beta \tag{4.44-7}$$

(note that ΔH is negative if one proceeds downhill in the direction of $+s$). The energy law then states:

$$\frac{7}{10}m(v_B^2 - v_A^2) = -mg\,\Delta H - f_r mg\cos\beta \cdot \Delta s \tag{4.44-8}$$

where f_r is the coefficient of rolling friction. Simplified this yields:

$$\frac{7}{10}\Delta(v^2) = -g\ \Delta H - f_r g \cos\beta\ \Delta s \tag{4.44-9}$$

Solving for f_r, one obtains:

$$f_r = \tan\beta - \frac{7}{10g\cos\beta}\ \frac{\Delta(v^2)}{\Delta s} \tag{4.44-10}$$

The difference quotient $\Delta v^2/\Delta s$ can be expressed as an acceleration. One has:

$$\frac{\Delta v^2}{\Delta s} = 2v\frac{\Delta v}{\Delta s} = 2v\frac{\mathrm{d}v}{\mathrm{d}t}\ \frac{\mathrm{d}t}{\mathrm{d}s} = 2a \tag{4.44-11}$$

with $a = \mathrm{d}v/\mathrm{d}t$. Thus, one can also write:

$$f_r = \tan\beta - \frac{7}{5g\cos\beta}a \tag{4.44-12}$$

It may also be useful to solve this for a. One has then:

$$a = \tfrac{5}{7}g(\sin\beta - f_r\cos\beta) \tag{4.44-13}$$

If there is no change in velocity (the rock keeps rolling) one has $a = 0$; hence:

$$f_r = \tan\beta \tag{4.44-14}$$

Thus it is found that the coefficient of rolling friction f_r is about equal to the tangent of the slope angle. The latter is equal to the angle of internal friction. The rather neat result is that the inclination of a scree slope is approximately given by the angle of internal friction φ of the material in question; that the coefficient of friction f_r for rolling blocks is of the order of $\tan\varphi$, but that the coefficient of friction f for debris slides (Sect.4.43) is considerably higher (Gerber and Scheidegger, 1973).

4.45. Protective measures

Protective measures on a scree slope must be designed so as to counter the three types of dangers discussed above: (a) internal instability and motion; (b) debris slides; and (c) rolling stones.

(a) With regard to stability it should be noted that in its natural state a scree slope is in a dynamical equilibrium. Anything affecting this equilibrium can have dangerous effects, e.g., the steepening of a slope by artificial or natural processes, or the thickening of the scree layer.

(b) The dynamics of a scree slide is determined by the energy conditions at the slope. If the slide is triggered by a rock fall from above, the energy input can be calculated, assuming that 3/4 of the energy is lost on impact, and using a coefficient of friction of $f = 0.6–0.9$ after this.

(c) Rolling stones are dangerous on slopes. The coefficient of (rolling) friction for such materials seems to be close to the tangent of the slope angle of natural slopes. Vegetation has been found to decrease the coefficient of friction. Therefore, on natural slopes, rocks can be expected to reach the bottom of a slope with the same velocity distribution they started with at the top.

Protective measures against the above phenomena include trenches (which must be built outside the area of the slope) to catch rocks and slides, stabilization structures on the slope and obstacles to increase the effective coefficient of friction. These must be dimensioned so that they can sustain the forces likely to be exerted by the anticipated events.

4.5. Surficial phenomena

4.51. General remarks

The phenomena occurring on scree slopes discussed in the foregoing section represent a special type of surficial process. Other types of surficial phenomena may also be involved in the origination of accidents of greater or lesser dimensions. A distinction is usually made between creep-type and slide-type phenomena.

If only the very top layer of a slope is involved in creep, one speaks of "skin" creep; if deeper layers are implicated as well, one speaks of "soil" creep. A further specialized type of phenomenon is niveal solifluction which affects the permanently frozen soil only.

Finally, surficial slips occur if only the vegetation cover of a slope is involved; thus, grass-sods often loosen and slip at higher altitudes.

4.52. Skin creep

Skin creep is a surficial phenomenon. It involves the zone of the seasonal temperature and moisture fluctuations; it rarely affects a depth of more than 3 m, and it is observed in all kinds of materials, viz. in clay, scree and rock. According to Penck (1929) skin creep is one of the main agents which transform a mountain range into a peneplain. Skin creep is spatially continuous and temporally in the mean (taken as average over the seasonal fluctuations) stationary.

The velocities of skin creep are surprisingly large. On shaley hillslopes in Colorado, Schumm (1967) measured movements of from 3 mm/year on a 3°-slope to 70 mm/year on a 40°-slope; in arid areas, Leopold et al. (1966) found movements reaching 5 mm/year. These motions can affect structures that are not properly anchored.

The mechanics of skin creep has been sought in some viscoplastic flow (Stroganov, 1961). The changing amounts of water in the ground also may have an effect (cf. Sect.6.32). Kirkby (1967) calculated theoretically a velocity profile from the assumption that the creep is due to alternate expansion and contraction of the soil. When the soil yields in this process, the soil particles are assumed to move in the direction of maximum stress. This yields the following velocity profile (v = velocity, z = vertical depth, β = slope angle):

$$v = \sin \beta \frac{(b + z) \exp(-bz)}{A^2} \qquad (4.52\text{-}1)$$

where A and b are constants; this velocity profile fits observed profiles rather well, where the velocity's rate of change is zero near the surface and at depth.

A more complicated soil-creep theory which is based on a stochastic motion of individual particles was attempted by Culling (1963).

Estimates of the forces acting in skin creep do not seem to have been undertaken yet. Therefore, one cannot give any quantitative design criteria for designing protective measures against skin creep. However, any measures to improve the stability of slopes such as good drainage, stabilizing structures etc., will certainly be helpful.

4.53. Surface creep

Skin creep implicates only those regions of a slope that are affected by temperature and moisture fluctuations. Creep, however, also occurs in deeper layers overlying a rocky core. This is called surface creep of the soil cover. Surface scree also affects rocks in which it may cause a downhill bending of the strata which is commonly called "outcrop curvature". This indicates that the downhill velocity of the creeping rock moves at an increasing rate when one is proceeding from within the slope towards its surface.

The creep of the *soil cover* over a rocky core was studied by Gerber and Scheidegger (1966) who showed that assuming the material as purely plastic qualitatively explains the observed phenomena. Finn and Emery (1970) calculated the surface creep with a finite element method on a computer using various viscoelastic flow laws. There is a well-known method for eliminating the time-dependence from the basic differential equation: a Laplace or Fourier transformation transforms the problem to an associated one in (static) elasticity theory (Alfrey, 1944), which can be solved by standard finite element techniques. Transforming the solution to original time yields the solution of the creep problem.

Finn and Emery (1970) used the method indicated for a calculation of the surficial creep of a long slope. Some examples were calculated, but a generally useful analysis of creep behaviour in non-dimensional variables, which could be easily applied to practical cases, still has not been developed. Therefore, the finite element calculations have to be made separately for each specific case.

As noted, surface creep *in rock* leads to outcrop curvature. The peculiar velocity profiles inferred from observations imply that the shearing stress is *increasing* towards the surface of the slope. This is mechanically difficult to understand, and the explanation found in the literature is that a superjacent cover of debris material has exerted a drag upon the rock. Ter-Stepanian (1965) has suggested some mechanical models of the process. However, outcrop curvature is also found in areas where there was never any debris cover over a rock.

Like skin creep, surface creep can also have catastrophic effects on the foundation of structures. It is still difficult to estimate the forces that are involved, so protective measures must rely on empiricism.

4.54. Solifluction

Substantial surficial movements on slopes can also be caused by the phenomenon of solifluction which is a process that occurs in frozen soil; it is caused by alternate freezing and thawing. In this it is closely associated with frost-heaving: frost-heaving occurs in a direction normal to the slope, settling upon thawing in the vertical direction so that the motion of the soil is parallel to the slope.

Many of the pertinent theories of solifluction have already been reviewed by the writer (Scheidegger, 1970a). In addition, Washburn (1970) deduced the following general principles from many investigations in Greenland.

(i) Solifluction in annually and permanently frozen soil is quantitatively different.

(ii) In permanently frozen soil the moisture content is the main factor causing the motion, which must be close to a critical value.

(iii) Solifluction is accompanied by retrogressive motions which diminish the resulting net amount of displacement caused by the ice in the ground, because the thawed layer has a tendency to settle back against the slope rather than vertically.

There is a relationship between solifluction rates and frost-heave rates. The latter are difficult to assess beforehand, because they depend to a large extent on the supply of water that is subject to freezing. Ladanyi (1972) has proposed a macroscopic theory by the assumption of a macroscopic creep law. In practical applications, the prevailing rates in the field have to be measured in order to design appropriate protective measures. These measurements are similar to those of any soil creep phenomenon. Techniques for making such measurements have been described by Williams (1957, 1959) and Washburn (1970).

4.55. Surficial slips

The surface layer of a slope may not only be subject to creep, but may also show discontinuous motions. There is a continuous transition from phenomena that involve only the actual surface of a slope to deep-seated, catastrophic landslides. The phenomenon of landslides will be discussed in Sect.4.7; we shall deal here solely with the most surficial slide phenomena such as like those represented by the slippage of the grass-cover of a mountain slope. A description of the phenomenon has been given by Patton (1956).

When the grass-cover of a slope is thoroughly soaked with water (due to a protracted rain period or the melting of snow) it may show signs of rupture, particularly below kinks in the profile. Large portions of the vegetation-cover may slide rapidly down the slope; they are sometimes broken into several pieces, which leave ugly scars on the landscape. These scars may remain without vegetation for a long period. If unchecked, a catastrophic denudation of a mountain side may result.

No mechanical models of this phenomenon appear to exist as yet.

4.6. Deep mass movements

4.61. Description

Catastrophic effects can also be caused by deep-seated creeping mass movements. The inexorable "closing" of valleys, which belongs in this category, has caused the destruction of many large bridges. Deep mass movements may occur in a variety of materials, ranging from soil to rock.

Mass creep in soil includes creep in all materials which can be treated by "soil mechanical" methods (cf. Sect.4.12). An example of this type of creep in overconsolidated clays occurred on the California coast (Gould, 1960) where a layered block 15 m thick and 330 m long was creeping along a pronounced gliding surface (slope angle $10^\circ - 12^\circ$) towards the sea with a velocity of from 2 cm/year to 10 cm/year. Similar instances of creep in clay have been reported by Bjerrum (1968) from southern Europe and the U.S.A. Mass creep in a clayey land slide mass was observed at the foundation of a railway bridge near Klosters in Switzerland (Haefeli et al., 1953) where maximum creep velocities were 37 mm/year. No gliding surface could be found in spite of drillings to a depth of 12 m. The movement was stationary. Large-scale (areas affected up to 40 km^2) and deep (50–150 m) "soil-creep" phenomena were also observed in the slates of the Hinterrhein valley in Switzerland (Jäckli, 1948). The slopes were generally about 25 : 100 and the largest movement was 26 cm/year. During the measurement period of 20 years, the motion decreased by 20%. The creep consisted of a combination of spatially continuous and discontinuous movements.

Rock creep is phenomenologically very similar to large-scale creep phenomena in soil, but it occurs in schist, limestone and even gneiss (Ampferer, 1939, 1940; Stiny, 1941; Zischinsky, 1967). It has the outward appearance of mass slumps. A typical example of rock creep is the Glunzerberg near Matrei in the Eastern Tyrol. The slope consists of phyllites and calcareous mica-schists; it has a height of 1,200 m and an average slope angle of 24°. In the uppermost regions of the slope, sliding surfaces were observed, but in the main part, no such surfaces exist. The movements are felt to a depth of about 200 m. Many slumps which started at the end of the glaciation have probably come to rest at the present time, but geodetic measurements show that there are also creeping "slumps" which are in almost stationary motion. Thus, Lanser (1967) found surface motions of 10 cm/year near Landeck (Tyrol) and Clar and Weiss (1965) found motions of 2.5–6.5 mm/day on the Millstätter Alpe (Carinthia). If these mass-

-creep phenomena are considered macroscopically, they can be regarded as continuous. Of course, they may be the result of many small-scale discontinuous displacements (Stiny, 1941, 1952; Horninger, 1958). Nevertheless, if the measurements are averaged over several years, a continuous and even stationary motion over many decades results.

4.62. Deep-seated soil creep

The theory of deep-seated mass creep in soil is based on the usual soil-mechanical principles. However, the usual soil-mechanical theory is based on limiting Rankine states only. For the description of creep it is necessary that time-dependence is considered. The creep characteristics of clays were studied by Bishop and Lovenbury (1969); the bearing of the long-term strength of clays on the depth-creep of slopes was analyzed by Goldstein and Ter-Stepanian (1957). A specific study of clay mobility in connection with large-scale creep phenomena at Palos Verdes in California was made by Kerr and Drew (1967).

Particularly in the last instance a slowly moving "slide" assumed catastrophic proportions. The properties of the clay in this instance follow a viscous flow law in which, however, the viscosity of the clay does not only depend on the shear rate, but also on the time during which the latter is applied. The pertinent values of the viscosity can only be obtained by laboratory tests of the material.

Once the material "constants" are known, finite-element methods exist to predict the behaviour of masses from similar materials to those developed for surficial creep (cf. Sect.4.53). Emery and Finn (1972) extended their method of making elasticity-analogs of creep problems to deep-seated creep in soil materials.

Regarding methods of control and prevention of large-scale deep-seated creep, it must again be stated that theoretical calculations of the motions and forces to be expected have not yet advanced very far. Accordingly, empirical methods have been employed. The general principles aiding the stabilization of a slope, such as good drainage, should always be applied. The installation of 35 caissons 6 m long failed to prevent the creep from advancing further at Palos Verdes (Kerr and Drew, 1967).

4.63. Rock mass creep

The basic phenomenology of rock mass creep has already been described in Sect.4.61. Accordingly, the phenomenon has the characteristics of a slow-moving "slump" (Zischinsky, 1969). Ampferer (1939, 1940) called the slow motion of the upper part of a slump "mountain fracture" while Stiny (1941) called the convex toe of the moving mass "valley closure". More recent descriptions have been given by Horninger (1958), Clar and Weiss (1965) and Zischinsky (1967, 1969). The velocity of rock mass creep may range from millimetres to metres per year (Altmann, 1958).

Sometimes, rock mass creep is attributed to tectonic stresses (Rybar, 1971) or to crustal movements, but we shall consider here solely the motion due to the action of gravity. At the beginning of the motion, there is a change in the stress field, such as the melting of the glaciers after the last ice age, so that the motion includes an adjustment to a change in the long-term stress field.

A mountain slope deformed by continuous mass creep in rock can be distinctly divided into three zones (see Fig.29).

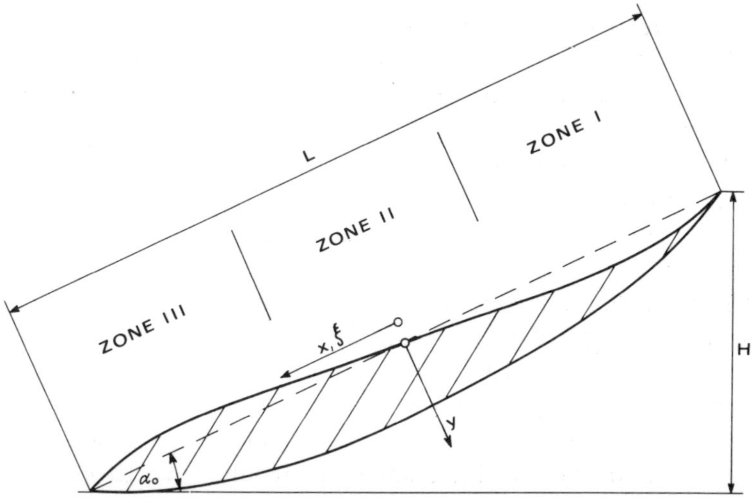

Fig. 29. Typical section through a slope deformed by spatially continuous mass creep. (After Brückl and Scheidegger, 1972.)

The uppermost part (zone I) represents Ampferer's "mountain fracture". The existence of tensional stresses can be inferred from the development of crevasses and the sliding on shear planes. During the deformation of the slope the surface is sinking below its original level. In the lowest part (zone III) we can infer the existence of compressional stresses because the surface is rising. This is the domain of "valley closure" described by Stiny (1941).

Between these two zones there exists a zone II in the middle part of the creeping slope where the longitudinal stresses are small or diminished compared with the shear stresses parallel to the slope surface. The deformation in zone II almost corresponds to simple shearing. The surface remains approximately at a constant level, although a diminution of the surface inclination can be observed. Because of this simple type of deformation, zone II is a suitable area to attempt a determination of rheological parameters. A corresponding analysis has been carried out by Brückl and Scheidegger (1972) who selected published reports on spatially continuous mass creep in rocks of the Eastern Alps. The information in these reports was then collated with four rheological models (generalized Newtonian fluid, Bingham body, logarithmic creep, Kelvin body) in order to test their applicability. The following conditions must be fulfilled by these rheological models.

(1) The creep velocity and the state of stress prevailing at the beginning of the creeping motion (at the end of the last ice age) are reasonably well known from the extrapolation of present-day observations.

(2) The state of stress at the present time and the total displacement since the beginning of the motion are known.

(3) The deformation must be spatially continuous and not a block motion.

(4) The deformation cannot be fully recoverable.

The main feature of these conditions is a tremendous decrease of the initial creep velocity from about 1 m/year 15,000 years ago to a few mm/year at the present time. In the same period, the decrease in shear stress was only 25%; therefore, the generalized Newtonian fluid and the Bingham body would have to be very sensitive to variations in shear stress. Logarithmic creep and the Kelvin body satisfy the above conditions by a viscosity increase

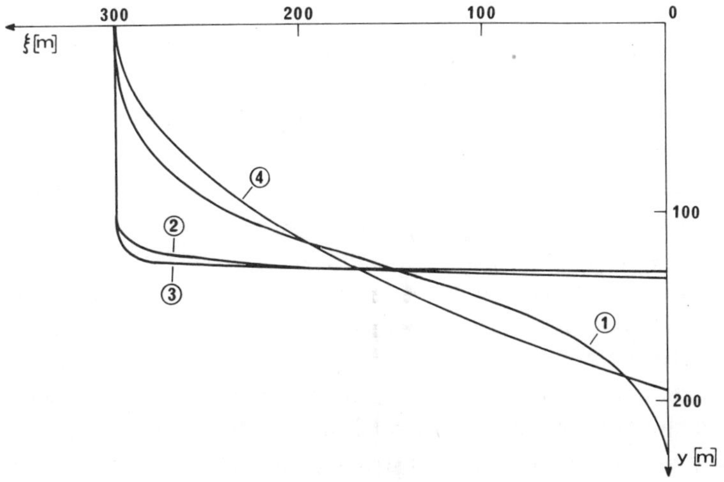

Fig. 30. Present displacement ξ as a function of depth y of the creeping rock. (After Brückl and Scheidegger, 1972.)
1 = observed displacement (Zischinsky, 1967); *2* = displacement calculated for a generalized Newtonian fluid; *3* = displacement calculated for a Bingham body; *4* = displacement calculated for logarithmic creep and a Kelvin body.

with time or with displacement. The possible solutions for the displacement ξ as a function of depth y are shown in Fig.30.

When fitting the parameters of the rheological models to the conditions which determine the course of creep in time, one obtains great differences in the spatial behaviour of creep. As we can see from Fig.30, the application of a generalized Newtonian fluid and of a Bingham body yields almost a block-type motion. This is contrary to the observed profile of creep which shows a continuous deformation of the whole creeping layer. Logarithmic creep and the Kelvin body fit the observational data much better.

In order to decide which is the most suitable of the two remaining rheological models, we can say that a Kelvin body yields fully recoverable creep. Regarding the rotation of the rock fragments within the creeping slope, it is difficult to imagine a recovery which would exceed a few per cent of the total deformation. Therefore, only the rheological model of logarithmic creep is appropriate to meet all the imposed conditions.

4.7. Landslides

4.71. General remarks

Landslides represent the rapid collapse of a slope. "Rapid", in this connection, means that the slide occurs in less than 15 min time. Naturally, premonitory movements or subsequent after-effects may be felt over much longer time intervals, but most of the motion is rapid; otherwise one is faced with such phenomena as earth-flows and mass creep which have been discussed in the previous sections of this book.

The beginning of a landslide occurs when the stability conditions of a slope are violated. In the present context, these stability conditions are not investigated; they have already been dealt with in Sect.4.2. We are concerned here with the mechanics of the actual slide.

The phenomenon of landslides has been described by many authors. Apart from the description of individual, single slides, reviews of the whole subject have been published by Heim (1932), Sharpe (1938), Cleaves (1961) and Záruba and Mencl (1969). A catalogue of landslides of the past 26 centuries has been compiled by Špurek (1972).

Many attempts have been made to classify landslides into characteristic types. Heim (1932) classified landslides principally according to the material in which they occur (loose materials, rock). It is of importance whether the sliding surface is self-generated or pre-existing. Kamenov et al. (1971) made a distinction according to the depth of the sliding surface. Záruba and Mencl (1969) also divided landslide phenomena according to the material involved and according to the depth of the sliding surface. Nemčok et al. (1972) made a classification according to the speed of the movement into creep, sliding, flow and fall. Of these, only the "sliding" motion would fall into our definition of landslides. The latter are further divided according to the depth and shape of the sliding surface.

A study of various case histories reveals that each landslide is, to some extent, an individual mechanical event, often quite different from all others. Nevertheless, a rough classification is necessary in order to survey the mechanical possibilities. Therefore, we shall first study entirely surficial phenomena (Sect.4.72), then discuss slides in soil-clay type materials (Sect.4.73) and, finally, discuss slides in rock (Sect.4.74). Integrating the various results, one can give some general guidelines for the prevention of slides (Sect. 4.75).

4.72. Surficial slides

Firstly, we will consider slides which affect only the surficial layers of a slope. It is to some extent arbitrary how thick a layer will be called "surficial". First included are the layers that are affected by the daily and seasonal temperature and moisture variations. However, these are not the only surficial layers. The term "surficial" is used when the thickness of the sliding material is comparatively small (less than 10%) in relation to the lateral extent of the sliding mass.

As noted in Sect.4.71, the initiation of the sliding motion is caused by a breakdown of static stability. The motion itself, then, is governed by the laws of gravitation and the laws of friction. The latter differ according to the type of the material that is involved in the slide.

In soil-clay type of materials, the mechanical behaviour of the substance is usually described by the cohesion c. The sliding mass will presumably be in an active Rankine state of instability. The final appearance of the slide is that of a series of slumps. The friction at the sliding surfaces is mainly governed by the cohesion c. An energy balance has to be developed for each individual slide based upon these factors.

According to the above, the energy balance of a mass m, subject internally only to a cohesion c, sliding downhill on a slope of slope angle β is:

$$\tfrac{1}{2}mv^2 = mg\ \Delta h - cA\ \frac{\Delta h}{\sin\beta} \qquad (4.72\text{-}1)$$

where Δh is the vertical distance through which the mass moves, v is the velocity (assumed as zero for $\Delta h = 0$) and A is the contact area. Setting:

$$m = \rho A \delta \tag{4.72-2}$$

where δ is the thickness and ρ the density of the sliding mass, one obtains:

$$\tfrac{1}{2}\rho A \delta v^2 = g\rho A \delta \Delta h - cA \frac{\Delta h}{\sin \beta} \tag{4.72-3}$$

$$v^2 = 2\left(g - \frac{c}{\delta \rho \sin \beta}\right) \Delta h \tag{4.72-4}$$

This shows that the slide continues for β = const. with increasing velocity, once it has started. The "slope" angle will generally be the inclination of some curved sliding surface; the latter will be one of the slip lines in a Rankine state of failure of a layer of material (cf. Fig.14). The slide will come to a rest when the more horizontal stretch of the sliding surface is reached. This causes the characteristic "slumping" appearance of these slides. Examples of this type of slide occurred on the banks of the Illinois river and in the hills of southern Alberta. The first case was described by Du Montelle et al. (1971) and the second case by Beaty (1972).

The opposite extreme is seen if the sliding material consists of rock. Again, the triggering of the slide is caused by the breakdown of stability (see Sect.4.2). The subsequent motion is again governed by the laws of gravity and of friction. Now, however, the friction is not given by the cohesion, but by the coefficient of dry friction f. The equation corresponding to eq.4.72-1 is therefore:

$$\tfrac{1}{2}mv^2 = mg\,\Delta h - mgf\,\Delta h \cot \beta \tag{4.72-5}$$

where the symbols have the same meaning as previously defined in this section. Hence:

$$v^2 = 2g(1 - f \cot \beta)\Delta h \tag{4.72-6}$$

One usually sets:

$$f = \tan \phi \tag{4.72-7}$$

where φ denotes the frictional angle (close to the angle of repose) of the sliding material and obtains then:

$$v^2 = 2g\left(1 - \frac{\tan\varphi}{\tan\beta}\right)\Delta h \qquad (4.72\text{-}8)$$

which shows that the material can slide only from a position of rest (the case of a slide after a rock fall where the mass does not start from rest, has already been discussed in Sect.4.43) if:

$$\frac{\tan\varphi}{\tan\beta} < 1 \qquad (4.72\text{-}9)$$

or:

$$\varphi < \beta \qquad (4.72\text{-}10)$$

If this condition is satisfied, the slide will proceed indefinitely downhill with an increasing velocity. A case of this type has been described by Vargas and Pichler (1957). The rock slide these two authors described came to a stop in the plane adjacent to the mountains.

4.73. Slides in soil and clay

We now turn our attention to slides in soil and clay which have a more deeply seated sliding surface than the surficial slides discussed above.

Again, the triggering of the slide and the establishment of the pertinent sliding surface is a question of statics. At the point where the dynamical problem has to be analyzed, the sliding surface represents a given boundary condition for the motion. In this connection, it is immaterial whether the sliding surface is the outcome of the failure process itself or whether it has been predetermined by the prevailing geological features, such as faults, unconformities or weak strata.

Therefore, in dynamic studies, the sliding surface is a predetermined quantity. The sliding mass is the material located between the sliding surface and the day-surface. In effect, little is known of the temporal evolution of such slides. Apart from slow and relatively steady creeping motions (which are of no concern in this chapter), the slides seem to occur in the form of one or more jerks. The evidence, however, is seen mostly after the phenomenon has occurred in the form of tear scars of greater or lesser dimensions. One case mentioned by Hutchinson (1965) seems to be better documented than most cases; Hutchinson mentioned that in the Vibstad slide of 22 February 1959 in Norway 10^6 m^3 of soil slid within "a few minutes" into the river Namsen below.

The motion itself is again determined by the laws of gravity and of friction at the sliding surface; to these must be added, under certain conditions, the laws of internal deformation. The slides that are easiest to understand are those where the sliding surface is cylindrical or spherical, so that the whole motion can be described by a single rotation. Then, the equations of motion can be formulated for that rotation. At the sliding surface, the shearing resistance is assumed to obey the usual Coulomb-Mohr law (eq.4.15-1), wherein the influence of the pore water may be introduced by using effective pressures (cf. Sect.6.32); the gravitational effect is described by the energy gain due to the vertical downward displacement of the centre of gravity of the sliding mass. By these means it is often possible to make an estimate of the various forces involved. It is clear from the laws involved (see Sect.4.72 for a discussion) that, once motion starts, it will accelerate until the geometrical conditions are altered by the motion until no driving forces are present. This explains the observed or inferred "jerks".

Cases where the slide can be represented essentially by a rigid rotation are common. The evidence is also to be seen on the surface, inasmuch as trees are bent backwards toward the slope without much internal strain in the sliding mass being visible.

More complicated cases occur when the sliding surface is one where substantial internal motions in the sliding mass are necessary to maintain contact. In these cases, the energy used for the internal deformations must be taken into consideration. Cases of this type do not occur infrequently; the sliding surface is chair-shaped with a steeply dipping upper part reaching to a nearly horizontal pre-existing geological contact surface. A dynamic analysis then requires assumptions regarding the energy required to bend the sliding masses into the shape of the sliding surface.

Finally, a peculiar phenomenon is represented by "quick clays". It has been found that some clays suddenly become quite fluid when a certain threshold is reached. This property is called "thixotropy". The fluidization can be induced by an increase of water content (after prolonged heavy rainfalls) or by the gradual leaching of salts from marine clays (Rosenqvist, 1953). The first of these cases occurred on May 4, 1970, in the town of St.Jean-Vianney (Quebec, Canada) (Anonymous, 1971); the second is a common enough occurrence in clay banks in Norway (Holmsen, 1953; Eide and Bjerrum, 1955; Bjerrum 1971b). When the salt content sinks below a certain level, the clay becomes thixotropic. The threshold values can be determined by experiments.

4.74. Catastrophic landslides

As was noted in Sect.4.43, the motion of a mass down a scree slope is governed by the laws of dry friction. For a constant coefficient of friction, the mass of the sliding mass, the average velocity and the local slope angle β are all eliminated from the final equations. The total vertical height of the slide divided by the total horizontal reach represents the average coefficient of friction f, provided the slide started from rest (cf. Sect.4.43).

It is a well known fact that for large, catastrophic landslides, the coefficient of friction becomes progressively smaller as the volume of the sliding mass increases, if one maintains a simple frictional model. Heim (1932) has already mentioned this fact and Scheller (1970) conjectured that the velocity of the slides increases with their volume. The coefficient of friction, then, would decrease with the increase of the sliding velocity.

Scheidegger (1973) analyzed the data from 33 catastrophic landslides that have been published. No correlations between the volumes (varying from 30,000 to $2 \cdot 10^{10}$ m^3) and velocities seem to exist. Heim (1932) noted that the velocity depends on the shape of the trajectory of the landslide, i.e., on the energy content available at each point of its path according to the frictional model. Therefore, Scheller's (1970) assumption of a correlation between velocity and volume does not seem to be supported by the data.

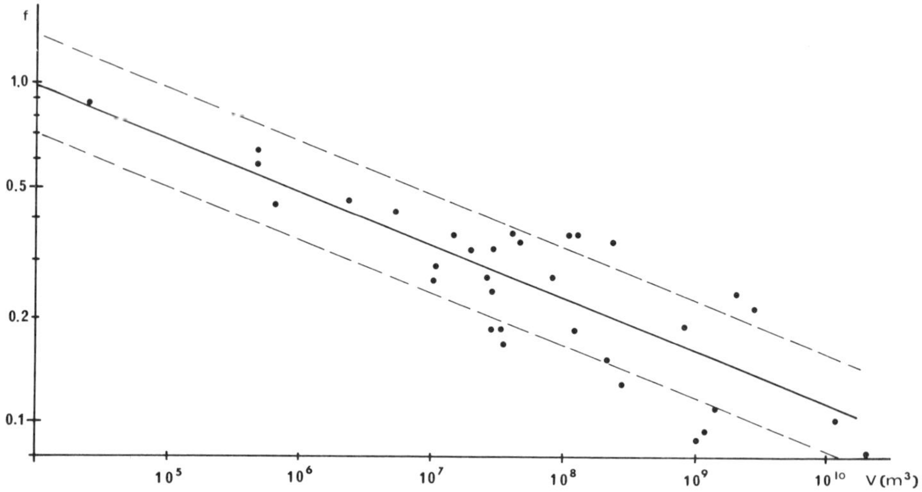

Fig. 31. Correlation between coefficient of friction f and volume V of large landslides. (After Scheidegger, 1973.)

The fact remains that there must be a *direct* correlation between sliding volume V and the coefficient of friction f. This correlation becomes evident if the slide data are plotted on a log-log scale (Fig.31). The correlation is then given by a straight line. Analytically, it is described by the equation:

$$\log f = a \log V + b \tag{4.74-1}$$

with:

$$a = -0.15\,666 \tag{4.74-2}$$

$$b = \quad 0.62\,419 \tag{4.74-3}$$

if V is measured in m^3 and the logarithms are to the basis 10. The correlation coeffient is 0.82. The dotted lines in Fig.31 represent the standard deviation of the data from the correlation line:

$$\sigma_{\log f} = 0.14298 \tag{4.74-4}$$

The correlation ends at the top by the value of $f = \tan \varphi$ where

φ is the angle of repose of the debris material involved. Most angles of repose are around 30°–40°, hence $\tan\varphi$ equals about 0.57–0.83. Therefore, the decrease of the coefficient of friction with volume becomes noticeable when the volume of the slide approaches 100,000 m^3 or so. The correlation line in Fig.31 enables one to predict the coefficient of friction that is obtained in an imminent landslide, if its volume can be estimated beforehand. With this coefficient of friction known, the reach and the velocity of the expected landslide can be estimated by means of the frictional model of Sect.4.43.

The phenomenological relation between the coefficient of friction of a landslide and its volume does not clarify the *mechanism* of the reduction of the friction. Why this occurs, is still a mystery.

Shreve (1959, 1968a, b) suggested that a cushion of entrapped air may form beneath the rock avalanche which may account for its great mobility. He (Shreve, 1968b) set up a model of a mass permeable to air that was supported by a trapped cushion of air. The mass can be supported only if the pressure gradient in the air at the bottom of the supported mass is greater than the gravity gradient in the mass:

$$\frac{\mathrm{d}p}{\mathrm{d}z} > \rho g \qquad (4.74\text{-}5)$$

where p denotes the pressure in the air, ρ the density of the mass, g the gravity acceleration and z the downward distance. The pressure gradient causes a flow of air through the mass which obeys Darcy's law (cf. eq.6.31-1):

$$q = \frac{k}{\mu}\frac{\mathrm{d}p}{\mathrm{d}z} \qquad (4.74\text{-}6)$$

where q is the filtration velocity, μ the viscosity and k the permeability of the mass to air. Elimination of $\mathrm{d}p/\mathrm{d}z$ yields:

$$q\mu/k > \rho g \qquad (4.74\text{-}7)$$

This flow will cause the trapped air cushion to become thinner.

Assuming the thickness of the trapped air to be initially 30 cm, Shreve calculated that no more than 1/10 of that thickness could be lost during the duration t of the landslide if the air cushion is not to lose its supporting quality. Making the assumptions that the duration t of a typical slide is 80 sec, the thickness of the slide 15 m, ρ equal to 2 g cm^{-3}, Shreve found:

$$q < 6 \cdot 10^{-2} \text{ cm sec}^{-1} \tag{4.74-8}$$

This leads to the following condition for the permeability k:

$$k \gtrsim 1 \text{ Darcy} \tag{4.74-9}$$

a value which does not appear as unreasonable. Therefore, Shreve's (1968b) model seems physically possible.

Nevertheless, other authors have proposed different models. Kent (1966) postulated a buoyant fluidization by the escape of air which was trapped between individual rock pieces throughout the thickness of the slide; thus he implied the validity of eq.4.74-5 for the whole sliding mass and did not restrict it to only the lower part of the mass. Goguel (personal communication) suggested that the heat caused by friction may vaporize the interstitial rock water, and that the steam formed would provide the fluidizing gas.

The phenomenology of catastrophic landslides definitely suggests a flow-type of phenomenon. Possibly, the process is analogous to the flow of turbidity currents at sea, a process which is also not well understood. Anyhow, the correlation between volume and coefficient of friction established here must be considered as fundamentally important in any final explanation of the phenomenon.

4.75. Prevention of slides

As was noted above, the initiation of a landslide has to be sought in every case in the breakdown of the stability conditions. Accordingly, measures for the prevention of landslides must first be directed towards ensuring that the stability conditions do not become violated.

The stability conditions for slopes have been discussed in Sect.4.23. A slope that has been stable for perhaps thousands of years may have a breakdown of the stability conditions in the following ways.

(i) The slope angle β increases gradually to the point where stability is no longer present. This can happen through slow geodynamical movements; however, it is caused more frequently by erosion at the foot of the slope (cf. Fig.23). This process occurs in the evolution (and collapse) of river banks and has been described by Kjaernsli and Simons (1962) in Norway, and Thomson (1970) in Alberta. The degradation of clay cliffs described by Hutchinson (1971) had a similar cause. Consequently, measures which prevent the steepening of a slope, e.g. the prevention of erosion at the foot of a river bank by corrective measures on the river, will be useful.

(ii) the instability can be caused by loading the slope with material which has been deposited artificially or naturally.

(iii) The instability can be caused by changes in the effective pressure. Any rise of the water pressure in the pores or joints contributes towards instabilization. Good drainage will therefore always enhance the stability of a slope. Drainage can also be aided by electro-osmosis, either by the instalment of electrodes or simply by short-circuiting different layers of the ground by the insertion of conducting rods.

For the stabilization of artificial slopes, several engineering means have been developed (earth butresses, shear keys etc.) which have been discussed in detail by Záruba and Mencl (1969). Some of them could also be used for the prevention of imminent *natural* slides.

Also of interest are premonitory effects. Before the occurrence of a landslide, creeping motions are often found to accelerate (Vajont; see Müller, 1968). Similarly, Goodman and Blake (1965) noted that there is a correlation between rock noise and the instability of a slope.

4.8. External flow phenomena

4.81. Introduction

In the present chapter we shall discuss a series of phenomena which could be characterized as external flows. In this instance, the word "flow" has to be understood in its widest sense: it includes plastic or logarithmic creep as well as the fast movements of a near-liquid.

In the phenomena treated here a distinction can be made between an external flowing material and the fixed substratum over which it flows. The flowing material undergoes large internal deformations, the substratum is essentially at rest. The flowing material may, in fact, be part of the original slope, or its source may be above the slope; in this aspect the distinction between the flowing material and deep mass movements or the landslide-type of phenomena may be a fine one; it lies essentially in the mechanical model that one makes for the phenomena discussed here: it is always assumed that one substance "flowing" over a "fixed" base can be recognized.

The phenomena will be further classified according to the speed with which they occur. We shall consider first slowly creeping flows, proceeding to progressively faster motions in the later sections of this chapter.

4.82. Slow earth- and mud-flows

First the slow flow of cohesive soil is considered. This phenomenon occurs in nature in the form of earth-flows and slow mudflows, these affect catastrophically any structure in their path. Occurrences of this type have been described by Zarubá and Mencl (1969; Handlová earth-flow, which had a total length of 1,800 m), by Crandell and Varnes (1960; Slumgullian flow in Colorado), by Moos (1953; earth-flow near Stoss in Switzerland), by Prior et al. (1971; Minnis north flow in northeastern Ireland) and by Hutchinson (1970). The last author has described a remarkable series of flows at the Beltringe Cliffs in northern Kent. Their average movement ranged up to 85 m/year, their discharges to 6,000 m^3/year.

Slow earth- and mud-flows have certain similarities with glaciers. They develop a tongue. Average inclinations of their surface reach from 7° to 12°. The flowing mass has sharp boundaries at its sides and base; its height tends to reach a certain limit when distance increases upward from the tongue. The flowing mass moves almost en bloc along the boundary surfaces (Hutchinson, 1970). Mean flow velocities reach from 4 m/year (Slumgullion) to about 600 m/year (Handlová).

Brückl and Scheidegger (1973) have attempted to give a mechanical explanation for the phenomena described above. This explanation is based on plasticity theory.

Accordingly, it is assumed that the flowing material is described by one single material constant; viz. the cohesion c. The shearing stress τ_s at the base, however, is described by two material constants: the cohesion c_s and the angle of internal friction φ_s:

$$\tau_s = c_s + \sigma_n \tan \varphi_s \leqslant c \tag{4.82-1}$$

In the above equation, σ_n is the effective normal stress. If β be the angle of inclination of the slip plane, and φ as well as β are small enough so that their sines can be set equal to the angles themselves and their cosines equal to 1, eq.4.82-1 can be written as follows:

$$\tau_s = c_s + \rho g h \left(1 - \frac{\rho_w h_w}{\rho h}\right) \varphi_s \tag{4.82-2}$$

where ρ denotes the density of the flowing material, h its height measured normal to the slip plane, ρ_w the density of the pore water and h_w the height of the hydraulic head of the pore water at the slip plane. The influence of the pore-water pressure expresses itself through the factor $\rho_w h_w / \rho h$.

For the further development of the theory, the problem is considered as two-dimensional. A Cartesian coordinate system is chosen with the origin in the slip plane, the x-axis pointing in the direction of the maximum slope and the y-axis directed upward and normal to the sliding plane. The most simple case is represented by an infinitely extended layer of flowing material with a constant inclination β and a constant thickness h. The solution for the normal stresses σ_x and σ_y and the shearing stress τ_{xy} can be found from Prandtl's solution of the plastostatic equations (cf. Hill, 1950, p.232):

$$\left.\begin{aligned} \sigma_x &= \rho g(h-y) \pm 2c_0\sqrt{1-(\tau_s/c)^2(1-y/h)^2} \\ \sigma_y &= \rho g(h-y) \\ \tau_{xy} &= \rho g(h-y)\beta \end{aligned}\right| \qquad (4.82\text{-}3)$$

The velocity component u in the x-direction (parallel to the slip plane) and the velocity component v in the y-direction (normal to the slip plane) are given by (Hill, 1950, p.232):

$$\begin{aligned} u &= u_0 \pm V\frac{x}{h} + \frac{2V}{\tau_s/c}\sqrt{1-\left(\frac{\tau_s}{c}\right)^2\left(1-\frac{y}{h}\right)^2} \\ v &= \mp Vy/h \end{aligned} \qquad (4.82\text{-}4)$$

The upper sign in eq.4.82-3 and eq.4.82-4 corresponds to the passive Rankine state and the lower sign to the active Rankine state. The quantities u_0 and V (velocity v_y for $y = h$) in eq.4.82-4 are constants of integration.

For an extension of the theory to layers with variable thickness, a relation which represents the equations of equilibrium in an integral form (cf. Collins, 1968, eq.7 on p.202) can be used. One obtains:

$$\tau_s = \rho g h\left(\beta - \frac{dh}{dx}\right) - \frac{d}{dx}\int_0^h (\sigma_x - \sigma_y)\, dy \qquad (4.82\text{-}5)$$

Assuming that $dh/dx \ll 1$, one can evaluate $(\sigma_x - \sigma_y)$ from eq.4.82-3 and insert the resulting expression into eq.4.83-5. Eliminating τ_s from eqs.4.82-2 and 4.82-5 one gets the following first-order differential equation:

$$\frac{dh}{dx}\left(h \pm \frac{kc}{\rho g}\right) - h\epsilon + \frac{c_s}{\rho g} = 0 \qquad (4.82\text{-}6)$$

where:

$$\epsilon = \beta\left(1 - \frac{\rho_w h_w}{\rho h}\right)\varphi_s \tag{4.82-7}$$

Furthermore, the quantity k is a factor which equals $\pi/2$ or 2 according to whether $\tau_s = c$ or $\tau_s \ll c$. If the flowing material is in a passive (active) Rankine state, the positive (negative) sign in the term within the brackets of eq.4.82-6 applies. As this term in eq.4.82-6 takes longitudinal stresses into account, all values between the two values of k given above are possible. However, the material in this case is not in a Rankine state and the flow can only occur en bloc.

If one assumes that ϵ = const, one can find the solutions for active and passive earth or mud flow tongues from eq.4.82-6. They are: (a) active tongue:

$$(1 - A)\ln\frac{1-\eta}{1-A} + \eta - A = \xi \tag{4.82-8}$$

where:

$$\left.\begin{aligned}
\xi &= \frac{\epsilon^2 \rho g}{c_s}\,x = \frac{\rho g}{c_s}\left[\beta - \left(1 - \frac{\rho_w h_w}{\rho h}\right)\varphi_s\right]^2 x \\
\eta &= \frac{\epsilon \rho g}{c_s}\,h = \frac{\rho g}{c_s}\left[\beta - \left(1 - \frac{\rho_w h_w}{\rho h}\right)\varphi_s\right] h \\
A &= \frac{2k\epsilon c}{c_s} = \frac{2kc}{c_s}\left[\beta - \left(1 - \frac{\rho_w h_w}{\rho h}\right)\varphi_s\right]
\end{aligned}\right|\tag{4.82-9}$$

(b) passive tongue:

$$(1 + A)\ln(1 - \eta) + \eta = \xi \tag{4.82-10}$$

where the symbols A, ξ, η have the same meaning as above. For the passive tongue, one has a thickness h of the flowing material which is zero at the origin of the coordinate system.

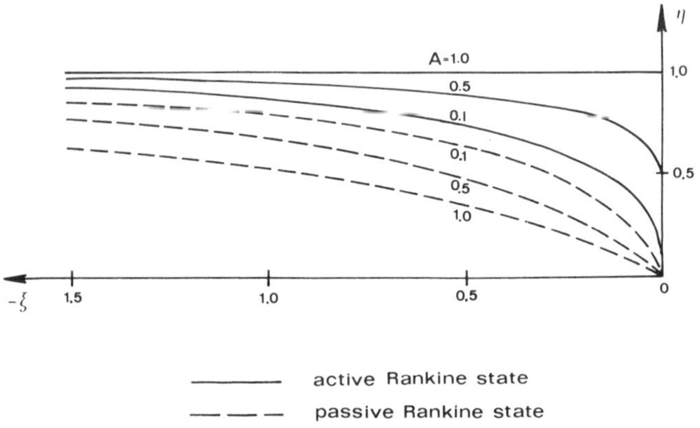

Fig. 32. Theoretical profiles of earth flow tongues for various values of the parameter A. (After Brückl and Scheidegger, 1973.)

Fig.32 shows the solutions represented by eqs.4.82-8 and 4.82-10 for various values of the parameter A. The thickness of the creeping mass tends to $h = c_s/(\epsilon \rho g)$ for $x \to \infty$ in either case. If the parameter A increases, the front of the tongue becomes steeper in the active state and flatter in the passive state.

According to Hutchinson and coworkers (Skempton and Hutchinson, 1969; Hutchinson and Bhandari, 1971) the causes of the motion of earth and slow mud flows are the accumulation of new material, an increase of the pore water pressure and erosion at the lower end of the tongue. These possible causes of the motion show that a Rankine state of motion may not necessarily develop in nature. However, according to the last two possibilities, an active Rankine state seems to be most probable.

A comparison of observed and theoretical profiles of earth and mud flow tongues is only possible if it is assumed that a Rankine state is completely developed. Brückl and Scheidegger (1973), assuming an *active* Rankine state, compared two observed cases with the theoretical profiles and the results obtained satisfactorily agreed.

The above model of an earth flow is based on Rankine states. However, different models have also been proposed. Cunningham (1972) considered an analogy with the flow of a viscous liquid and

Smalley (1972) considered a thixotropy-type of effect to be responsible for the transition from a stationary to a flowing state.

4.83. Fast mud- and debris-flows

In contrast to the slow "mud"-flows considered in the last section, there also exist *rapid* flows of dense mixtures of mud, debris and water. These cannot be entirely separated from flash floods (cf. Chapter 6), inasmuch as there is a gradual transition from debris-flows to water-flows, depending on the density of the flowing mass.

The phenomenon of the mud-debris-water flows in Alpine territories was described by Stiny (1910) in a presentation that is still pertinent. A mudflow can be compared with a lava stream running down a gulley. Its surface is strongly convex, the "current" may be up to 30 m in thickness; mean velocities may surpass 10 m/sec and have been observed to attain 19.5, locally 30 m/sec (Aulitzky, 1970). The motion may occur in spurts, local blockages may temporarily bring it entirely to a halt. The downstream velocity is greatest in the center of the cross-section, some distance below the free surface of the flow. A mud-flow follows a pre-existing gulley, but not as readily as a water flood. The density of a mud-flow is high (up to 2.5 g/cm^3; it may correspond to up to an 80% content of solids by volume) so that rocks appear to flow in it. Its total volume may range from $5 \cdot 10^3$ to $4.75 \cdot 10^5 m^3$.

The mechanical explanation of fast mud-flows has generally been based on river-flow formulas (cf. Sect.6.26). Thus, Gol'din and Lyubashevskiy (1966) took as fundamental relation:

$$v = C\sqrt{RS} \tag{4.83-1}$$

where v is the mean downhill velocity, R the hydraulic radius (essentially the thickness of the flow), S the slope of the bed and C Manning's factor equal to:

$$C = \frac{1}{n} R^{1/6} \tag{4.83-2}$$

with n the roughness factor (cf. Scheidegger, 1970).

From experimental investigations with artificial flows, the following empirical relations were found (Gol'din and Lyubashevskiy, 1966), giving the velocity in m/sec:

$$v = 3.15\, H^{1/6}\, d^{1/3} \quad \text{and} \quad v = 5.15\, H^{2/3}\, S^{1/4} \tag{4.83-3}$$

where H(m) denotes the thickness of the flow and d(m) the average grain diameter of the debris in a *debris*-flow. In a *mud*-flow one has:

$$v = 3.15\, H^{1/6}\, d^{1/3} \sqrt{\frac{\rho_s - \rho_m}{\rho_m}} \tag{4.83-4}$$

where ρ_s is the density of the rock debris and ρ_m the density of the flow mixture.

The above formulas are based upon an analogy of mud-flows with riverbed processes. It is not yet certain whether these formulas should actually be used for mud- and debris-flows with a high content of solids (Zeller, 1972). In fact, it has been observed that debris-flows can be brought to a halt (at least temporarily) by an obstacle as small as the remnants of a snow avalanche. This leads to the conjecture that the moving debris mass is in a state of fluidization owing to the internal water pressure within it. If the latter drops below the fluidization threshold, the mass immediately comes to a halt.

For fluidized solids of high concentration in a liquid, Bagnold (1954) postulated the existence of a dispersive stress. In a fluidized mass, two flow regimes are possible; these are termed "inertial" and "macroviscous" regimes. In the inertial flow regime, which is characterized by a high shearing velocity, the dispersive stress is caused by the collisions between individual grains; in the macroviscous regime, which is characterized by a low shearing velocity, the dispersive stress is caused by an interaction between the grains and the fluid. A comparison of the numerical values required by Bagnold's theory with those found in debris-flows shows that Bagnold's theory may be invoked as an explanation of the mechanism of debris-flows.

In addition to the motion of a mud-flow, the condition for its origination is also of importance: i.e., the question of the development of a simple flash flood into a highly dangerous mud-flow. Only qualitative answers can be given to this question (Zeller, 1972). Critical flow velocities and a sufficient supply of debris material must be present. This problem has not yet been solved. An empirical correlation exists between rainfall and probability of a mud-flow (Rapp, 1963; Aulitzky, 1968; Prior et al., 1970; Hofmann, 1971; Williams and Guy, 1971; Kojan et al., 1972). However, no sure predictions of an impending disaster can be made on this basis from rainfall data as of yet. More hopeful are correlations between *flow* and debris load (Wehrmann, 1950; Hofmann, 1969). It appears that the usual river flow formulas are no longer applicable, inasmuch as bed load transport and mud-flows appear to be entirely different phenomena (Hofmann, 1971). Thus, the mechanics of mud-flows is not yet understood.

Chapter 5. SNOW AND ICE CATASTROPHES

5.1. Introduction

Many accidents and catastrophes are caused by the action of snow and ice. The best known of these are due to avalanches which are a constant menace in mountainous regions whenever substantial amounts of snow are on the ground. Such avalanches may affect a few lone skiers, may block highways and railroads, or may bury whole buildings and parts of villages. Ice-falls from surging glaciers may have even more dramatic effects: 4,000 human lives were lost due to an ice-fall from a glacier on Mt. Huascaran (Peru) on January 10, 1962, which subsequently travelled as a giant slide and, after melting, as a water-squall down a valley, destroyed nine villages.

The physical problems connected with snow and ice catastrophes are very similar to those encountered in accidents which occur on "slopes". Therefore, the general methods of soil and rock mechanics have applications in snow and ice mechanics, respectively. Some of the basic laws of the stress–strain behaviour of snow and ice are different from those that govern the behaviour of soil and rock, but nevertheless the problems encountered in snow and ice mechanics are entirely analogous to those encountered in soil and rock mechanics.

Therefore, it will first be necessary to establish the general stress–strain relationships and strength laws for snow and ice; then problems of the dynamics of large masses of such materials can be investigated.

5.2. Physics of snow and ice

5.21. Freezing water

Water is a most unusual substance, because its liquid state is denser than its solid state: when water freezes, it expands. The basic crystallographic structure of frozen water is hexagonal. When water vapour freezes in the atmosphere, it usually forms small crystals which are precipitated as snow. The snowflakes have a great variety of forms; usually they are in the shape of small hexagonal plates with complex dendritic branches (Nakaya, 1954), which are sometimes in the shape of hexagonal prisms or needles.

As the snow falls to the ground, it forms a loose, porous layer. The sunlight and percolating water slowly transforms the loose snow into a more solid and dense material. Recrystallization may occur until the aspect of solid ice is presented.

To describe the physics of snow and ice is not a simple matter. Snow exists in many different varieties, and ice may exhibit various types of polycrystalline structures as well as several high-pressure phases (Fletcher, 1970). For practical applications, however, only the macroscopic properties are of interest. These will be reviewed below.

5.22. Mechanical properties of snow

Snow is a granular solid. When stressed, it responds with irreversible deformations, firstly in a continuous manner (creep, compaction) and, secondly, by exhibiting a phenomenon akin to failure.

Continuous deformations

The continuous creep of snow under shearing stresses can generally be described by a viscous model (Kartashov, 1966; Haefeli, 1967). There does not seem to be any stress threshold present below which snow does not flow. The stress–strain relation for snow is linear for small stresses. However, when a certain stress limit is exceeded, the strain rate rises more rapidly than would correspond with a linear relationship. It is, however, difficult to

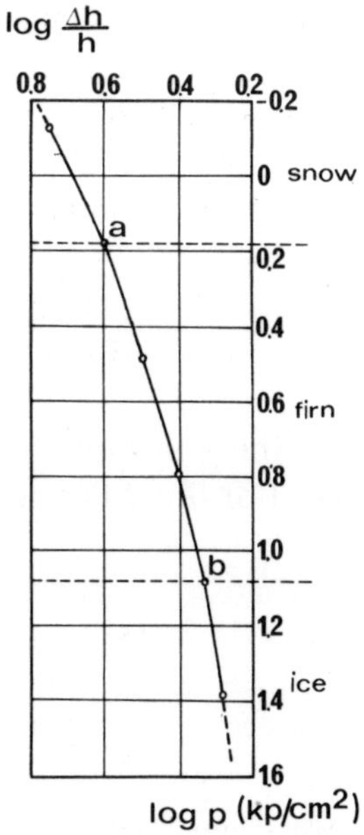

Fig. 33. Ultimate compaction Δh of a snow layer of thickness h under a vertical pressure p. (After Kartashov, 1966.)

give values for the "viscosity" of "snow", since these widely vary, depending on the consistency of the snow. Moreover, when creeping, a snow-cover becomes compacted and slowly transforms itself into increasingly dense states until it gains the appearance of *ice*. It is not logical to consider the "creep" of snow without taking the compaction and the transformation of snow into ice into consideration. For the compaction of snow under pressure alone, several relationships have been proposed. According to Kartashov (1966):

$$p = A(\Delta h/h)^{\alpha} \tag{5.22-1}$$

where p is the vertical pressure, h is the original thickness of a snow cover and Δh is the final compaction caused by the pressure p. A and α are constants. The result from a typical set of measurements is shown in Fig.33.

Failure in snow

Failure in snow can be described using the usual Coulomb–

Mohr criterion, with a zero angle of internal friction (Ballard and McGraw, 1966). The "strength" of snow is used to express failure.

Because the quality of snow varies greatly, it must be characterized by a further quantity. This quantity is its porosity P. Then, the strength σ_{crit} of snow is given by a formula of the following type:

$$\sigma_{crit} = A(1 - P/P_0) \tag{5.22-2}$$

Theoretically, A is the strength of ice ($P = 0$), the constant P_0 depends on the type of strength that is required. For the shearing strength, Fuchs (1949) gives:

$$A = 13.8 \text{ kp/cm}^2 \tag{5.22-3}$$

$$P_0 = 0.490 \tag{5.22-4}$$

for tensile strength (Butkovich; cf. Ballard and McGraw, 1966):

$$A = 28.3 \text{ kp/cm}^2 \tag{5.22-5}$$

$$P_0 = 0.544 \tag{5.22-6}$$

5.23. Mechanical properties of ice

When studying the mechanical properties of ice, it should be noted that ice occurs in a variety of forms. Two of these forms are of particular importance: columnar-grained (S) ice and granular (T) ice (Michel and Ramseier, 1969). There are two types of columnar-grained ice, denoted as S1 and S2. In S1 the planes of easy slip are perpendicular, in S2 these planes are parallel to the direction of freezing. The T-ice has random crystallographic orientation of the grains.

The mechanical behaviour of ice is then very dependent on temperature, the rate of loading, etc. Although ice is, in many ways, a solid, it also exhibits creep properties. The stress–strain-rate relationship is nonlinear. It can formally be represented as a viscous-flow law of the type:

$$\tau = 2\eta\dot{\epsilon} \tag{5.23-1}$$

(where τ is the shearing stress and ϵ the (tensorial) shearing strain rate) with a "viscosity" η that depends on the shearing stress. Lliboutry (1970) has reviewed the possibilities that have been suggested in the literature:

$1/\eta = 0.296\ \tau^{4.2}$ (Glen)

$1/\eta = 0.25\ \tau^2$ (Haefeli)

$1/\eta = 0.036 + 0.128\ \tau^{4.5}$ (Meier)

$1/\eta = 0.036 + 0.080\ \tau^2 + 0.053\ \tau^4$ (Lliboutry)

The most commonly used form of the flow law is that of Glen (1955).

The above flow laws are nonlinear, which greatly complicates their mathematical application. Fortunately, the linear-flow law of plasticity theory is quite adequate in many cases when the yield strength is taken as equal to 1 bar (Paterson, 1969, p.88). The flow laws for ice are the basis for the explanation of the motion of glaciers.

Ice also behaves as a solid. Its linear compressibility has been measured (Gow and Williamson, 1972) as approximately 4 mbar^{-1}.

When a certain tensile or shearing stress is exceeded, ice will fracture. The fracture of ice under an unaxial stress was studied by Hawkes and Mellor (1972); the failure of ice in general was investigated by Gold (1970). Gold reported that the failure of ice occurred with the formation of cracks, which begins when a constant compressive stress exceeds about 6 kp/cm^2. In a tensile stress state, the critical load is slightly lower. These cracks are originally uniformly distributed spatially; only when the stress exceeds about 12 kp/cm^2 do they tend to concentrate in zones parallel to the planes of maximum shear, thereby causing fractures. The crack formation also depends on the strain rate; therefore, the failure behaviour of ice not only depends on the stress state, but also on the rate of loading, particularly when under compression. It is difficult to interpret tests of "strength" of ice, because the "creep-type" yielding invariably interacts with the "fracture-type" behaviour. Thus, for low strain rates, a very low yield "strength" is found, because the ice begins to creep immediately. Only at higher strain rates, above 10^{-6} sec^{-1} (where creep is understood to occur in any case) is a critical "fracture"-type behaviour separable from the whole deformation pattern. In this range, the tensile strength of ice is found to be around 10 bars; the compressive strength increases (linearly versus the logarithm of the strain rate) from about 10 bars at creep rates of 10^{-6} sec^{-1} to 80 bars at 10^{-3} sec^{-1} where it appears to even out (Hawkes and Mellor, 1972).

5.24. The transformation of snow to ice

As noted above, snow is gradually transformed into a more solid condition until the consistency of ice is reached. This process is a complicated thermodynamic phenomenon.

Freshly fallen snow at low temperature in calm weather may have a density of 0.01 to 0.03 g cm^{-3}; in comparison glacier ice has a density of 0.85–0.91 g cm^{-3}. Between these limits, all intermediate values are possible. Snow can be transformed to ice in two different ways, depending on whether or not water is present in the pores of the snow.

The transformation of dry snow into ice involves the mutual displacement of crystals and internal deformations. Lliboutry (1971) developed a theory explaining the gradual closing of pores in a snow mass which is based on Glen's law (cf. Sect.5.23) of ice creep. The speed with which the holes close, depends on the temperature.

When Lliboutry's formulas are applied to actual cases, it appears (Petit, 1971) that the observations of densities with depth in snow fields cannot, in general, be explained only by the closure of pores due to the overburden pressure: ice occurs at a much higher level in a snow mass than predicted by the theory. Hence, the percolating melt water must have a profound influence on the formation of ice from snow. Again, there may be several zones which have water content in a snow mass: one zone in which the pores are only partly filled with water, another where they are completely filled (soaked zone). The conditions are very similar to those encountered in groundwater studies; the dividing line between the partially and completely saturated zones corresponds to the groundwater table. Such a water table exists also for snow masses.

Snow itself is a very poor conductor of heat, but the water percolating through a snow mass is a major agent in transferring energy. The local freezing and thawing, the heat transfer and subsequent transformation of snow to ice is a very complicated process in which many phenomena participate. It is not the task of a treatise on catastrophes to discuss this in detail, and therefore, the intersted reader is referred to the pertinent literature cited above.

5.3. Snow

5.31. Phenomenology of avalanches

Avalanches have been known for a very long time. The earliest written records of avalanches can be traced to the Romans, who mentioned that Hannibal encountered them when he was crossing the Alps (cf. Silius Italicus, ca.90).

Avalanches represent simply the collapse of a mass of snow which may engulf man and beast and destroy property. In this, snow avalanches are similar to surficial landslides. In order to study their dynamic aspects, it is first necessary to distinguish between the various phenomenologically different classes of avalanches. To this date no generally accepted classification scheme has been devised: a number of proposals were presented during the International Symposium on Scientific Aspects of Snow and Ice Avalanches at Davos in 1965. One of the most widely used classifications is probably that due to Haefeli as amended by De Quervain (1966). This classification table, in a somewhat simplified form, is shown in Table VII.

The various phenomenological avalanche-types are not neces-

TABLE VII

Types of avalanches
(Simplified from De Quervain, 1966)

Criterion	Characteristics and name	
A. Type of breakaway	start from a point loose-snow avalanche	start from a line slab avalanche
B. Position of sliding surface	within snow-cover (top strata only involved) surface-layer avalanche	whole snow-cover sliding full-depth avalanches
C. State of humidity	dry snow dry-snow avalanche	wet snow wet-snow avalanche
D. Form of track	open, flat unconfined avalanche	in a channel channeled avalanche
E. Form of movement	through the air airborne powder avalanche	on ground flowing avalanche
F. Trigger factor	internal spontaneous avalanche	external triggered (natural or artificial) avalanche

sarily different in their dynamics. It is therefore not possible to discuss them one by one. Rather, it will be expedient to review the various dynamical possibilities in a manner similar to the chapter on landslides. The initiation of an avalanche must be sought in a breakdown of static stability of the snow-cover; the subsequent collapse is a dynamic process.

5.32. The stability of snow-covers

The stability of a snow-cover can, in principle, be calculated by the usual methods of stability analysis of a continuous material. However, the consistency of a snow-cover changes constantly during its slow metamorphosis. Thus, a stability analysis which does not include the compaction and metamorphosis of a snow-layer would not be pertinent.

The combined effect of the various factors can be expressed by the introduction of a "creep vector" for every original point of a

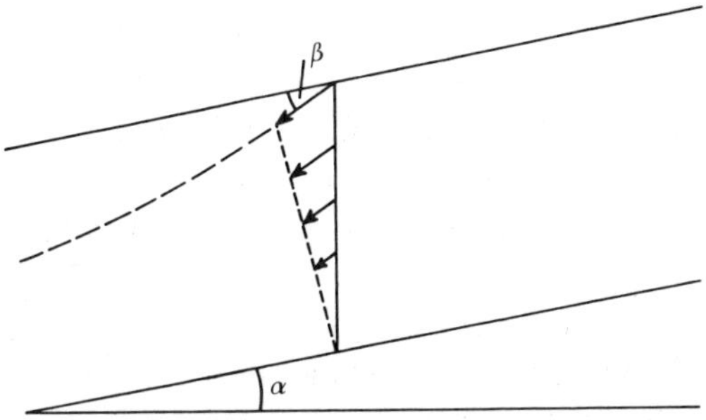

Fig. 34. Creep vectors in a snow-cover.

creeping material; the creep vector represents the displacement experienced by a point during a given, constant time interval. In a snow-cover lying on a plane-inclined slope it appears (Haefeli, 1967) that the creep vectors are all parallel and fill a triangle (Fig.34). The angle which the creep vector forms with the slope surface, is called the "creep angle". The path which a point describes during its creeping motion is called a "creep curve". If the creep angle is known (from empirical measurements) the direction of the principal stresses in a snow-cover can be determined. The method of this determination is illustrated in Fig.35. Let P be the point at which the creep vector $\boldsymbol{v}$ has been determined. The normal to this creep vector intersects the ground surface at M, around

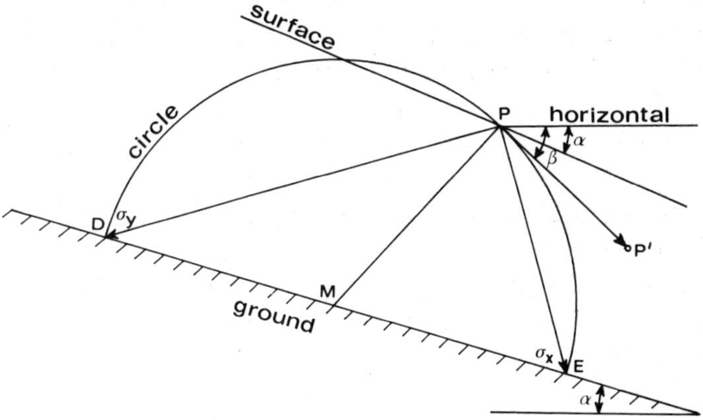

Fig. 35. Construction of principal stress directions σ_x and σ_y in a snow-cover.

which a half-circle *DPE* is drawn. The angle *DPE* is a right one. When the point *P* moves to a new position *P′* along **v**, the angle *DP′E* must again be a right one. This is possible only if *DP* and *EP* are principal stress directions. With progressing compaction, the creep angle decreases; hence a self-stabilization of a new snow-cover may occur.

As a simple criterion for a stability analysis one can set the condition that failure occurs when the stresses parallel to the slope are tensile or exceed the values given in Sect.5.22. Haefeli's slope angle analysis was generalized by Perla (1972).

The degree of stability of a snow-cover may also be expressed by means of a "stability index" (Roch, 1966):

$$s = \tau_s / \tau \qquad (5.32\text{-}1)$$

where τ_s is the shearing strength in a plane parallel to the slope and τ is the shearing stress in the same plane due to the weight of the snow. These quantities must be measured or estimated for particular snow conditions. It is found that avalanches may begin when $s = 4$, and, in fact, always do when $s < 1$.

A more detailed analysis of the state of stress in fallen snow has been reported by Brown et al. (1972). An avalanche is released when the following conditions occur (Perla and Lachapelle, 1970): (1) shear failure at the bottom of a sliding slab (against snow or ground); (2) tensile failure at the top of the slab; (3) shear failure at the sides of the slab; (4) compression failure at the bottom of the slab. The model is clearly quite complicated; the resulting equations are hard to use in practice. The phenomenon of buckling in a snow slab at the bottom was further studied by Lang et al. (1973).

To estimate the probability of an avalanche occurring, the use of the stability analysis of Roch (1966) is probably the most practical procedure. Empirical values for the strength of the snow as it is present on the ground and for its thickness give an idea of the stability of the snow.

5.33. Dynamics of avalanches

Dynamical types of avalanches

In the phenomenological classification of avalanches, a whole series of different types was introduced. However, amongst all these types, only two are dynamically different (Shoda, 1966): surface-flow avalanches and airborne-flow avalanches. The names of these two avalanche-types are self-explanatory.

Surface-flow avalanches

For surface-flow avalanches, a dynamical model has been developed by Salm (1966). It is based on the idea that a mass m of snow is sliding downhill on a uniformly inclined plane at slope angle β. The acting forces are the weight W and a frictional resistance R.

The equation of motion is then:

$$m\dot{v} = \frac{\rho_{\text{snow}} - \rho_{\text{air}}}{\rho_{\text{snow}}} mg \sin \beta - R \tag{5.33-1}$$

where v is the downslope velocity and the density (ρ) factor allows for buoyancy forces in air. The friction R depends on the prevailing resistance mechanism. Considering the possibility of dry friction, viscous flow and turbulent resistance, Salm (1966) wrote:

$$R = R_0 + R_1 v + R_2 v^2 \tag{5.33-2}$$

which is, incidentally, the beginning of a Taylor series in v.

The dry friction term R may be set as equal to:

$$R_0 = L\rho f g(A \cos \beta + S) \frac{\rho_{\text{snow}} - \rho_{\text{air}}}{\rho_{\text{snow}}} \tag{5.33-3}$$

where ρ is the density of the snow, f a friction coefficient, S is the side-friction factor per unit length (dimension: an area), A the cross-sectional area and L the length (measured downslope) of the sliding mass. The side friction S is proportional to the square of the depth of the sliding mass:

$$S = \lambda_\alpha d^2 \tag{5.33-4}$$

where λ_α is a coefficient.

The term R_1 corresponds to viscous "friction":

$$R_1 = \eta \frac{1}{\delta} PL \tag{5.33-5}$$

where η is the (effective) viscosity, δ the thickness of the viscous boundary layer and and P the contact perimeter of the sliding mass.

Finally, the term R_2 corresponds to turbulent resistances, corresponding to Chézy's equation for rivers:

$$R_2 = LP\rho g/k^2 \tag{5.33-6}$$

where k is a roughness factor.

Evidently, if the expressions for R_0, R_1 and R_2 are used in the equation of motion, a differential equation for v is obtained. The equation, in fact, is no different from the basic equation of flow of water in a channel.

For surface-flow avalanches the turbulence term can be neglected. The solution of the differential equation, then, depends on the extent of the boundary layer. For slab avalanches, where the sliding is concentrated in a very thin layer of thickness δ (effectively, the frictional resistance increases linearly with velocity) the result is:

$$v = \left[\sin\beta - f\left(\cos\beta + \frac{S}{A}\right)\right] \frac{\rho_{\text{snow}} - \rho_{\text{air}}}{\rho_{\text{snow}}} \frac{A\delta}{P} \left[1 - \exp\left(\frac{-\eta P}{A\rho\delta} t\right)\right] \tag{5.33-7}$$

where the expression in front of the second square bracket (see Eq.5.33-7) represents the maximum possible velocity. Measurements have yielded $\eta/\delta = 0.0475$ kp sec/(dm)^3. For a typical avalanche with $A/P = 1$ m, $\rho = 0.2$ g/cm^3 and:

$$H \equiv \left[\sin\beta - f\left(\cos\beta + \frac{S}{A}\right)\right] \frac{\rho_{\text{snow}} - \rho_{\text{air}}}{\rho_{\text{snow}}} = 0.3 \tag{5.33-8}$$

the ultimate velocity $v = 1{,}263$ m/sec. However, the neglected turbulence may prevent this value from being actually attained.

If the boundary layer is assumed to encompass the entire snow-layer, the influence of the viscosity is negligible. Thus, the solution is:

$$v = \sqrt{\left[\sin\beta - f(\cos\beta + \frac{S}{A}\right] \frac{\rho_{\text{snow}} - \rho_{\text{air}}}{\rho_{\text{snow}}} \frac{A}{P} k^2}$$

$$\times \tanh \frac{Pg\sqrt{HAk^2/P}}{Ak^2} t \qquad (5.33\text{-}9)$$

Furthermore, one has $k^2 \approx 500$. One can, therefore, describe this type of avalanche by dry friction alone (only the first term need be taken into account in the Taylor expansion of the hyperbolic tangent).

When the term S/A is neglected the usual frictional model for any slide is obtained (cf. Sect.4.43). The reach of the slide is given by formula 4.43-6, which, in turn, makes it possible empirically to calculate coefficients of friction f. Novikov (1970) has found a correlation between f and the density ρ for new-snow avalanches (density ρ between 0.12 and 0.22 g/cm^3) which is:

$$f = A\rho \qquad (5.33\text{-}10)$$

where $A = 2.75$ cm^{3}g^{-1}, yielding a range for f of from 0.25 to 0.6. For drift and metamorphosed snow avalanches f varies erratically from 0.55 to 0.76.

The above model allows one to calculate run-out distances of avalanches. One basically has the model of a mass sliding downhill, dry friction being the only resisting force. The calculations are similar to those made for the analogous landslides. The pertinent coefficient of friction, or better, the pertinent value for the expression in eq.5.33-8 has to be used.

One may wonder whether the above description of the mechanism of avalanches is entirely adequate. In the case of landslides, it has been found that the effective coefficient of friction f does not depend on the velocity (as implied by eq.5.33-2), but on the *volume* of the sliding mass. The mechanism by which this dependence is effected is, at present, not yet known. Nevertheless, one may surmise that avalanches are not very different from sliding soil masses and therefore, a similar correlation between them may exist. However, no attempts to find the correlation appear to have been undertaken as yet.

Airborne-flow avalanches

Salm (1966) also included in his model airborne powder avalanches. The friction term R_0 would then be zero, whereby the turbulent resistance will then be the most important factor. In fact, one obtains in this fashion a simple model of a turbidity current.

It is reasonable to assume that the mechanism of airborne powder-snow avalanches is indeed similar to that of turbidity currents. The general mechanics of a turbidity current, as far as it is known (cf. Sect.6.52), would therefore explain the mechanism of airborne flow avalanches (Allen, 1971a).

5.34. Destructive forces in snow avalanches

The frictional model of avalanche motion, which was introduced in the last section, is a macroscopic model. This model does not allow one to calculate the distribution of the stresses *within* an avalanche which would make it possible to estimate the impact forces. A model by means of which such an estimate can be achieved was proposed by Voellmy (1955), who compared the internal structure of a moving avalanche with that of a flowing river. The only additional assumption that Voellmy made concerns a relation between the density of the flowing snow and the prevailing pressure. The velocity, the height of the flowing layer and the pressure exerted upon obstacles can then immediately be deduced from the usual river-flow equations.

In this way Voellmy arrived at a set of 10 equations which, although they contain a series of phenomenological coefficients, allow one to estimate the destructive forces of avalanches upon structures. Accordingly, the pressure head H on impact is:

$$H = h' + \frac{v^2}{2g}\,[1 - (v_{\rm n}/v)^2\,(1 - \sin\beta)] \qquad (5.34\text{-}1)$$

where h' is the thickness of the flowing layer, v its mean velocity, v_n its impact velocity and β the angle between the velocity and the surface which is being hit. The pressure exerted on the obstacle is then:

$$p = \rho g(H - h^*) \tag{5.34-2}$$

where ρ is the density of the flowing layer at the point of impact and h^* the height above the bottom of the avalanche. Empirically, one has for h' values ranging from the original thickness h of the snow-cover to 2.6 h.

Typically it is found that impact pressures range from very small values to 7,300 kp/m^2. In addition to a frontal pressure, there is also an upward drag (similar to a buoyancy effect) which tends to tip over houses etc. The total upward force P_V upon a house front of area A is approximately:

$$P_V = cpA \tag{5.34-3}$$

where c is an empirical coefficient equal to about 0.3. Similar formulas have been deduced for the forces on a horizontal, covered roof, for thin poles etc. The reader is referred to the original paper of Voellmy (1955) for the details.

5.35. Statistics and forecasting of avalanches

In any given area, some winters produce more avalanches than others. It is clear, therefore, that avalanche activity is closely connected with the meteorological conditions, as are, for instance, floods and hurricanes. One would think, therefore, that a similar approach as that to the statistics of floods could be applied to the statistics of avalanches. Perhaps one could count the number of avalanches that occur every year in a given region and find the probability that a certain number is exceeded in a given time interval, just as this is done with river floods. In this fashion an estimation of the avalanche risk in an area could be obtained.

The writer is not aware of this approach having yet been taken, although it would appear to be reasonably promising. Crecy

(1966) has begun to collect statistical data of avalanche frequency, but did not proceed to an evaluation of them in terms of a stochastic model.

Most evaluations of avalanche risks pertain to an analysis of the particular conditions at a specific locality at a given time. In this, considerable success has been achieved, inasmuch as from the measurements on the prevailing conditions of a snow-cover its stability can be estimated. For the local forecasting of avalanches this requires a synthesis of observations and theory; the requirements have been reviewed by Lachapelle (1966). Primarily the methods of stability analysis outlined in Sect.5.32 will have to be applied.

5.36. Protective measures against avalanches

In areas where avalanches are frequent, the problem arises of protecting life and property from their destructive effects. The protective measures that can be taken are the following.

(A) An accurate assessment of the stability of an existing snow-cover can be made. Then, if the snow-cover in a region is found to be unstable, a general order that this region be avoided by tourists, skiers etc., can be issued. A menacing avalanche can be triggered at a known time by mortar shots. Attempts to stabilize an unstable snow-cover can be made by trying to influence the snow metamorphism by chemical agents (Lachapelle and Stillman, 1966). For this purpose, aldehydes (benzaldehyde and N-heptaldehyde) have been used with moderate success.

(B) Attempts can be made to erect structures which would provide a stabilization in areas where avalanches commonly originate in winter: the idea is to prevent avalanches by building permanent supporting structures. Generally, scattered supports have been used; these, however, have not been very successful (Voellmy, 1955; Frutiger, 1966) in preventing avalanches, unless they are built as continuous walls and are very dense indeed.

(C) Measures can be taken to deflect a moving avalanche from reaching a sensitive area. The most successful protective structures are wedges placed immediately in front of the endangered area; the obstacle will cause the avalanche to rise in height; therefore, the angle of the wedge must be small enough and the structure high enough to prevent the avalanche flowing over it. The various design criteria have been discussed by Voellmy (1955).

(D) The endangered buildings, etc., can be built to withstand the forces caused by an avalanche. In this connection, the upward drag has to be particularly noted so that a house will be sufficiently anchored to prevent it being toppled (cf. Sect.5.34).

5.4. Glacier catastrophes

5.41. Phenomenology

Catastrophes caused by glaciers usually originate with the break-off of a large chunk of ice which, in turn, triggers some ice or mud avalanches. Some remarkable incidents of this type have occurred in recent times: on January 10, 1962 some $3 \cdot 10^6$ m^3 of ice broke off the higher slopes of Mt. Huascaran in Peru, triggering an ice and mud avalanche which killed some 4,000 people (Morales, 1966). The accident was repeated on a larger scale (this time triggered by an earthquake) on May 31, 1970 whereby some 70,000 people lost their lives (Welsch and Kinzl, 1970). A similar accident occurred near Mattmark in Switzerland on August 30, 1965, killing 88 persons (Brockamp et al., 1967). In all these cases an ice mass became unstable, either by a sudden advance ("surge") of the glacier in question, or due to an earthquake. The ice mass then travelled down a valley, in some cases liquefying itself by melting, and dragging much additional debris along.

The problem of forecasting such accidents is apparently based on two phenomena: the sudden "surging" of a glacier which may greatly increase the volume of the ice mass near its tongue, and the stability of that ice mass itself, possibly when affected by earthquake accelerations. The motion of the avalanche, once it is detached, follows the usual laws of avalanche dynamics, whereby the melting of ice which may be particularly severe in tropical regions, represents an additional factor to be reckoned with.

5.42. The normal motion of glaciers

Before treating catastrophes caused by glaciers, i.e., before dealing with their "abnormal" behaviour, it will be necessary to have a general idea about their "normal", common behaviour. We are not, of course, concerned here with a comprehensive review of glaciology, and therefore present only the most general aspects of glacier motion. Some of the basic aspects of glacier motion may be found in the writer's own monograph on geomorphology (Scheidegger, 1970a) as well as in a specific treatise on the physics of glaciers by Paterson (1969).

Thus, there are two modes of motion of a glacier. Firstly, there is internal creep within the ice according to a plasticity-type or Glen-type law (see Sect.5.23). Second, there is the sliding of the glacier over its bed. The longitudinal movement of a glacier is caused by these two effects, each contributing about one-half of the total motion (Kamb, 1970). Normal speeds of glaciers are 30–150 m/year in the Alps, 700–1,300 m/year in the Himalayas and 1,000–3,000 m/year in Greenland.

The internal creep of a glacier has been particularly investigated by Nye (1951) and by John and Sugden (1971). It explains many of the observations of the internal stress state in a glacier, including the origination of crevasses and similar features.

The sliding motion of glaciers also contains much pseudo-creep. It seems that the sliding occurs primarily when ice in front of bed-roughnesses is pressure-melted, and with the regelation of the melt-water behind the obstacles (Weertman, 1957; Lliboutry, 1964a). In general, the sliding is then expressed *phenomenologically* by a macroscopic friction model. The calculation of the macroscopic friction factor from a microscopic bed model is subject to much argument (Weertman, 1967a, 1971; Kamb, 1970; Nye, 1970).

In the macroscopic sliding of glaciers, the role of the water contained in the crevasses and in the region between the glacier and its bed is of great importance. Studies of this subject have been reported by Weertman (1971b, 1972) and by Nye and Mae (1972). For phenomenological purposes, it is generally sufficient to employ the usual soil-mechanical relationships in which the effective pressure is introduced as the difference between the external and the fluid pressure. Examples of this procedure will be given in connection with stability calculations (Sect.5.44).

Because of the mathematical difficulties in dealing with glacier motion, particularly if Glen's law is used as the basic rheological condition for ice (see Sect.5.23), model experiments also have been undertaken. A particularly successful effort in modeling the flow and sliding of glaciers has been reported by Fleisher and Sales (1972) who used modeling plaster to represent the ice. Structures very similar to those encountered in ice were obtained.

5.43. Surges of glaciers

...atastrophes finding their origin in glacier phenomena are often ... with so-called "surges". Such "surges" represent ...pid advances of a glacier which may lead to an unstable ...tion of the glacier ice and subsequently cause avalanches. ... avalanche of Huascaran in 1962 (Morales, 1966) as well as ...tmark accident of 1965 (Brockamp et al., 1967) may have ...aused by a glacial surge.

...ges of glaciers do not always lead to catastrophes; they have ... observed, e.g. on hundreds of glaciers in North America (Meier and Post, 1969; Post, 1969) and in glaciers of Greenland (Rutishauser, 1971). Evidence of ancient glacier surges was found in the Swiss Alps by Röthlishberger (1969) and in Alaska by Post (1972). According to Paterson (1969) glaciers move during surges at velocities which are about 100 times the normal velocity. Post (1960) found a movement of about 6.5 km in not more than 9 months on the Muldrow glacier in Alaska, yielding a velocity of 20 m/day.

The mechanism of the initiation of glacier surges is still not completely understood. There are some general rules for the mass balance of a glacier (Kasser, 1970), but these are not sufficient to explain instabilities that occur suddenly.

The possible causes for the latter were enumerated by Robin (1969): Accordingly, the following causes may be effective: (a) stress instabilities; (b) temperature instabilities; (c) bed-friction instabilities.

Stress instabilities

Stress instabilities in a layer of ice can be caused by recrystallization of individual portions of the glacier under strain (Jonas and Müller, 1969). Robin (1969) and Robin and Barnes (1969) showed that stress concentrations in ice can propagate themselves as waves and can thereby produce unusually high rates of flow, an idea which has also been followed up by Palmer and Nye (1972). A modification of the stress-instability theory of the initiation of glacier surges is also represented by the ice-dam theory of Nielsen (1968, 1969), in which the build-up of an ice-dam with subsequent collapse through crumbling is thought to be the cause of the sudden surge.

Temperature instabilities

The warming of a cold glacier also could contribute to its sudden, rapid advance (Lliboutry, 1969). However, it has been found that both temperate and cold glaciers can surge (Harrison, 1972). Therefore it is unlikely that a temperature instability is a primary cause of the surging of a glacier.

Bed-friction instabilities

The most likely cause of the surge of a glacier is a sudden decrease in the bed-friction. Several mechanisms that could produce such a decrease have been discussed in the literature. Weertman (1962) has claimed that catastrophic sliding of a glacier is possible when the thickness of the water layer at the bottom of a glacier exceeds the size of the bed obstacles, a theory which was amplified later (Weertman, 1969). Similarly, Lliboutry (1964b) assumed that the friction of a glacier bed can be reduced inordinately by detachment (supercavitation) of the ice. Using a strictly viscous-flow law for the ice, Campbell and Rasmussen (1969) found that a reduction of the friction at the bed by a mere 5% yields a surge. Brückl (1972) also showed that a small improvement of the sliding conditions will yield a surge.

The existence of numerous different theories of glacier surges is an indication that the problem has not been solved to date. Therefore, it is almost impossible to make estimates of the probability that a glacier will undergo a big surge at any particular time. With present-day knowledge predictions of glacier surges cannot be made from mechanical theory. The only available way to make such predictions, therefore, is an investigation of the past behaviour of a glacier, assuming that it will surge again in the future if it has been subject to surges in the past. A statistical analysis of the available data will give one at least some idea of what is to be expected.

5.44. Stability of an ice mass

The actual detachment of an ice mass from a glacier or ice field is a problem of static stability. During the motion of a glacier, this stability may disappear. According to Haefeli (1966b) there are basically two modes by which an ice mass can fail: when the glacier is pushing forward, an ice wedge may fall from the front of it, or a whole slab of ice may become unstable on its bed.

Wedge failure

A glacier pushing over the last point of its frontal support where, say, the inclined bed drops off into a steep wall, either forms an overhang (if the ice slides over the bed) or a progressively steeper frontal slope (if the ice does not slide over the bed). The principal vertical stress is equal to the overburden weight of the ice (Haefeli, 1963), and the usual methods of analysis can be used to estimate the stability of the ice-"slope" bank. However, only relatively small masses of ice are released at a time in this way.

Slab failure

The great ice avalanches (Huascaran, Altels, Mattmark) represent instances of slab failures. Again, the prevailing stresses can be calculated by the usual soil-mechanical equations. Thus, the shearing stress τ_s at the bed is:

$$\tau_s = c + (\sigma - p) \tan \varphi \tag{5.44-1}$$

where p is the water pressure and φ the (effective) angle of friction. For an ice slab, the cohesion c at the bed can probably be set equal to zero. For equilibrium, one must have (Haefeli, 1966b):

$$\sigma_z = \rho g l(\sin \psi - \cos \psi \tan \varphi) + p \tan \varphi l/d \tag{5.44-2}$$

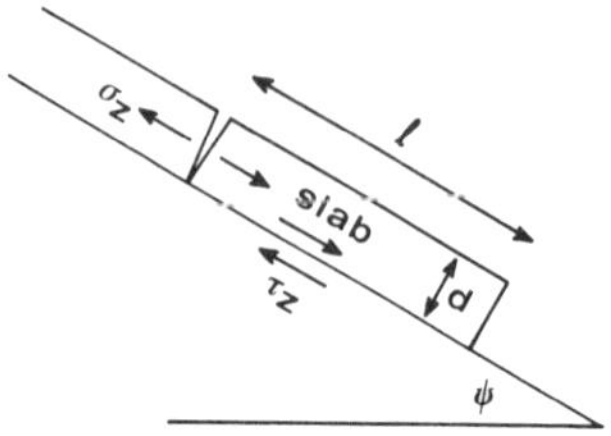

Fig. 36. Definition-sketch for failure in an ice slab.

where ρ is the density of the ice and the meaning of the other symbols becomes evident upon an inspection of the definition sketch in Fig.36. A test using observed values for l, d and ψ yields that the stability is greatly affected by the water pressure p and the angle of friction φ at the bed of the glacier which must be inferred *a posteriori*. When σ_z exceeds the tensile strength of ice (see Sect.5.23), a break-off of the slab will occur.

If the tensile strength of the ice is taken as zero (crevasses!) then the equilibrium condition (5.44-2) becomes:

$$\tan\varphi \geqslant \frac{\sin\psi}{\cos\psi - (p/\rho g d)} \tag{5.44-3}$$

Because of the *a priori* uncertainty in the values of p and φ, eq.5.44-2 cannot easily be used for an estimate of the stability of a particular glacier. It does yield, however, an assessment of the possible causes of an ice avalanche that has occurred: abnormal water pressure p, or a decrease in the thickness d of the glacier during periods of ice retreat.

Eq.5.44-2 does not take external accelerations into account. Therefore, the stability of an ice tongue may be affected by the large accelerations occurring during an earthquake: in this respect it is similar to the stability of a slope. According to D'Alembert's principle, such an acceleration a is equivalent to a force equal to ma where m is the mass of the ice slab. Thus, the component of the earthquake acceleration parallel to the bed $a_{//}$ simply effects an addition of $-a_{//}\rho$ to σ_z, the component normal to the bed a_b effects a change in the contribution to σ_z equal to $-a_b \rho d \tan\varphi$. It is clear that the stability of the slab can be greatly affected thereby. This explains the Huascaran avalanche of 1970 which was triggered by an earthquake.

5.45. The motion of ice avalanches

When an ice mass has detached itself from a glacier, it proceeds downhill initially like a debris slide. However, during its motion its consistency may change inasmuch as it is transformed into a rapid mud-flow (presumably by the melting of the ice by friction or meteorological heat) which continuously collects additional material. Thus the volume of the moving mass may increase tremendously during the flow.

Unfortunately, actual data on ice avalanches are few. Information gleaned from the literature is presented in Table VIII. Based on the frictional model of sliding masses (cf. eq.4.43-6) an average effective coefficient of friction f was calculated from the data. It is obvious that there is no correlation between it and the volume of the slide; the reverse occurred with large landslides (cf. Sect.4.74). The value of f for the Altels avalanche agrees with those found for landslides (cf. Fig.31) but in the other cases f is much lower. One may infer from this observation that one limit case in an ice avalanche is represented by the same mechanism as in large landslides. However, in many instances the friction is lower than in a

ABLE VIII

ata on ice avalanches

cation d date	Total height (m)	Distance (m)	f	Initial volume (m^3)	Final volume (m^3)	Reference
tels Sep., 1895	1,440	3,255	0.44	$4.5 \cdot 10^6$		Haefeli, 1966
uascaran Jan., 1962	4,000	16,000	0.25	$2.5-3.0 \cdot 10^6$	$13 \cdot 10^6$	Morales, 1966
attmark Aug., 1965	400	1,400	0.28	$2.0 \cdot 10^6$		Brockamp et al., 1967
uascaran May, 1970	4,000	16,000	0.25	$2.5 \cdot 10^7$	$5 \cdot 10^7$	Welsch and Kinzl, 1970

landslide of corresponding volume. Since transformations of ice avalanches into fairly liquid mud-flows have actually been observed, this is not at all surprising. Another limit case may occur when the shattering of the ice actually produces a snow avalanche in which f may be around 0.25. Thus, for the calculation of the reach of an impending ice avalanche by means of the frictional models, the corresponding ranges for f should be taken.

Chapter 6. WATER CATASTROPHES

6.1. Introduction

6.11. Phenomenological classification

Water may be involved in several ways in catastrophic events. Probably the catastrophes involving water that are first thought of will be floods in river basins which are caused by severe storms. The prediction of the frequency and of the severity of floods is a very important problem indeed in a study of the physical aspects of natural catastrophes. Related to floods caused by storms are floods caused by the sudden emptying of a lake. The damaging effects of these events are related to each other in the respect that it is *surface-water* which is the active agent.

In land areas, not all the water moves across the surface of the earth; a substantial portion is found in the ground as *ground-water*. Ground-water can also cause disastrous effects: it can cause a loss of stability in a slope in which it is contained, its removal can cause consolidation or its influx can cause sand to become quicksand. It can even trigger such geodynamic effects as earthquakes.

Most of the water on the earth is contained in large bodies: lakes and oceans. The boundary lines between these bodies of water and the land, i.e., the shores and coasts, are subject to special types of problems: catastrophic effects may be caused by sea-level change and by large flood waves, such as caused by tsunamis.

Finally, catastrophic events may also occur below the water-line, at the *ocean-bottom.* Subaqueous slides may have similar dimensions as landslides, triggering turbidity currents which can have far-reaching effects upon man's works.

6.12. The hydrological cycle

Any discussion of the geomorphological effects of water must begin with the *hydrological cycle*. Water from oceans and lakes evaporates, is carried aloft and finally condenses in the sky to form clouds. Eventually, it is precipitated back to the earth as rain, snow or dew. Some of the water then seeps into the soil and rock to proceed as ground-water, the other part forming surface-water. Finally, the water returns again to the ocean.

The above description of the hydrological cycle is a greatly simplified presentation. Evaporation does not only occur from oceans and lakes, but also from plants (evapotranspiration), from rivers and directly from the soil; so in the over all hydrological cycle there are also small, partial cycles.

For the global hydrological balance one has (see e.g. De Wiest, 1965, p.38ff):

precipitation in oceans	81.9 cm/annum
precipitation on land	67.4 cm/annum
evaporation from oceans	92.1 cm/annum
evaporation from land	42.0 cm/annum

A multiplication of these values by the respective surface areas (oceans 70.8%, land 29.2% of the earth's surface) does not yield a balance; thus there is a net *run-off* of water from the land to the sea; on land areas, it averages approximately 25 cm/annum.

It is the run-off water which causes geomorphic effects. This is the water that flows on and through the ground, thereby affecting the landscape. The water on the earth's surface is termed surface-water, the water flowing in the ground is termed ground-water. The surface-water collected in drainage basins (catchment areas) flows into rivers and streams which are susceptible to various amounts of flow. The ground-water can cause the formation of caves, affect the stability of mountain sides and trigger geodynamic phenomena.

Furthermore, the ocean (and other large bodies of surface-water) can interact with the land on coasts and shore lines. This leads to rather special phenomena of great interest. Finally, morphological effects can also occur within large bodies of water on the land- (or rather, "sea"-) scape below its surface.

We shall discuss these phenomena in turn below.

6.2. Surface-water

6.21. The hydrograph

In the study of surface-water, various hydrological quantities can be measured, such as water-level, through-put, etc. The representation of the course in time of any such variable is called a "hydrograph".

The most common hydrographs refer to stream-flow Q or the water-level depth in a river. The latter is usually expressed in terms of hydraulic radius R defined as the cross-sectional area A of the river divided by the wetted perimeter. For large rivers, R becomes equal to the mean depth. The quantities Q and R are related by the Manning formula (see, e.g., Scheidegger, 1970a):

$$Q = \frac{1}{n} A\, R^{2/3}\, S^{1/2} \qquad (6.21\text{-}1)$$

where Q is in m^3/sec, A in m^2, R in m, S is the slope gradient (a pure number), and n is a roughness factor which can be estimated empirically for typical channels. Barnes (1967) has produced a remarkable collection of type-cases for the estimation of n, using colour pictures to demonstrate each type.

It is thus possible to convert water-level measurements at a point in a river into hydrographs of stream-flow Q versus t. The type of curve that results when using this method is sketched in Fig.37. It is one of the aims of hydrological science to "explain" such hydrographs.

In the hydrograph of stream-flow at a particular point in a river, the meteorological events in the whole drainage basin (i.e., the total area contributing to the river-flow) are of significance. Of particular interest is the hydrograph of a flood-flow caused by a rain-storm (or a rain-storm *plus* snow-melt) over the drainage area.

A flood hydrograph (cf. Fig.37) consists of a rising limb (starting at point A in Fig.37), it has a crest and then a recession limb. In the hydrograph, several types of flow components are com-

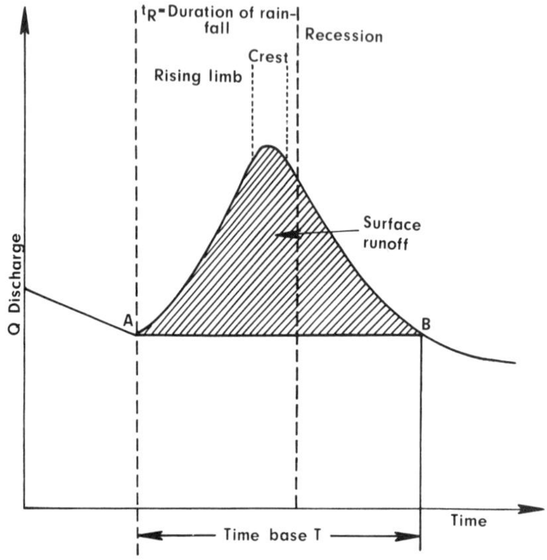

Fig. 37. Schematic sketch of a hydrograph.

bined. In particular, the total run-off is commonly divided into two parts: direct run-off and base flow. The direct run-off is that part which is evident in the stream immediately after the rainfall (or snow-melt). The base flow is the sustained run-off; it contains ground-water and delayed subsurface flow. The division is somewhat arbitrary, and accordingly there are various techniques to separate hydrographs into their components. A simple method is shown in Fig.37, where the base flow is represented by a straight horizontal line which begins at the inflection point A and reaching the hydrograph again at point B.

It would be advantageous to be able to use hydrographs from known storms to predict the hydrographs for expected storms. A method of doing this, based upon a superposition principle, has been initiated by Sherman (1932) and has been widely applied and extended since (Chow, 1964).

Accordingly, one makes the following fundamental assumptions:

(a) The time-base of a direct surface run-off hydrograph for a given drainage basin for a constant rainfall rate storm depends only on the duration of the rainfall, not on its volume.

(b) The ordinates of the direct surface run-off hydrographs for a given drainage basin for constant rainfall rate storms of a given (fixed) duration are proportional to that rate.

(c) Hydrographs from different storms in a given drainage basin can be linearly superposed.

Because of the above assumptions, one can first of all construct "unit hydrographs" representing hydrographs for a unit precipitation rate for storms of various fixed durations. However, one can do even more: because of the superposition principle, the hydrograph of a drainage basin can be expressed in terms of a convolution integral:

$$Q(t) = \int_0^t x(\tau)\, u(t-\tau)\, \mathrm{d}\tau \qquad (6.21\text{-}2)$$

where x is the rainfall rate and $u(t)$ is the response function. The latter is equal to the hydrograph $Q(t)$ obtained for $x(\tau) = \delta(\tau)$, i.e., for a unit rainfall occurring in an infinitesimally small time interval at $t = 0$. It is called the instantaneous unit hydrograph:

$$Q(t)_{\,x(\tau)=\delta(\tau)} = \int_0^t \delta(\tau)\, u(t-\tau)\, \mathrm{d}\tau = u(t) \qquad (6.21\text{-}3)$$

The basic function characterizing the response of a drainage basin to rainfall is thus the instantaneous unit hydrograph. In principle, it can be obtained from any hydrograph $Q(t)$ for a known rainfall rate $x(\tau)$, for a given drainage basin by solving the convolution integral eq.6.21-2 for the function $u(t)$. Theoretically, this can easily be done by using Laplace transforms. In practice, this may be difficult to perform, therefore many approximate procedures have been proposed in the literature. A review of the possible methods has been given by Chow (1964, pp.14–25ff).

The validity of the superposition assumption may, however, be doubtful (Nemec, 1972, p.248). Each fall of rain has an individual pattern and the volume of direct run-off influences the hydrograph not only as a proportionality factor, but also in some nonlinear fashion. Thus, nonlinear modifications of the linear unit hydrograph theory have been suggested (Kraijenhoff van de Leur,

1966). Nevertheless, the response of suitable drainage basins (up to 5,000 km^2 area) does seem to correspond to the unit hydrograph theory at least approximately.

The estimation of the possible peak flood in an area, beginning with an estimate of the possible peak precipitation as input into the response function, is based upon this fact. Practical cases have been discussed by Alexander (1972) and by White and Reich (1970).

6.22. Statistical analysis of river hydrographs

The hydrographs introduced in the last section have been treated as single responses to rainfall. Of great interest is the study of the statistical properties of the long-time hydrograph of a river, particularly with a view to prediction of uncommon events (floods and droughts).

The deterministic correlation between rainfall and floods leads one to expect a connection between the statistical properties of the hydrograph of a river and the statistics of the meteorological elements referring to the corresponding drainage basin. Reich (1970a,b) compared some flood series with rainfall extremes. In a thorough study of Pennsylvania floods (Reich, 1970a), he found no usable correlation between the extreme value statistics of rainfall and floods. A better correlation was obtained between flood statistics and certain basin parameters for basins in a given region (Reich, 1970b; Rogers, 1972). Duckstein et al. (1972) demonstrated the difficulty of using equi-spaced data for correlation purposes, except under special conditions. Eagleson (1972) investigated the dynamic chain between rainfalls and the floods they cause; in principle, it is then possible to relate the statistics of one with those of the other. This relation is, however, by no means simple.

The difficulties with the deterministic method suggest that the hydrographs should be investigated per se. One can first analyse the statistical properties of the flow $Q(t)$ (treated as a time series) per se, and then try to explain these properties by proposing a statistical *model*. If the model is correct, predictions can be made. Unfortunately, this procedure has not been brought to an unequivocal conclusion, therefore many statistical models are current in the literature.

The most direct approach to developing a model is by determining the correlation statistics of $Q(t)$, represented, for instance, as a (time) series of daily flow readings taken at appropriate gaging stations in a river (e.g., Hall, 1972). Veitch and Shephard (1971) applied this method to the river Murray in Australia, and found that the hydrograph could be predicted approximately for 12 days from its correlation statistics. Needless to say, this is not sufficient for the treatment of floods.

Hydrological time series can also be analysed in a more sophisticated fashion, by comparing them with possible models. Matalas (1967) has given a general review of the properties of such hydrologic time series. A characteristic feature is that hydrologic events are not independent. For independent events one would, in a presumed stationary state (in the long run) have "Brownian" conditions (Markov chain with no autocorrelation). Denoting the random variable by $Q(t)$, one can introduce a cumulative random variable:

$$q(t) = \int_{\text{const}}^{t} Q(t)\,\mathrm{d}t \qquad (6.22\text{-}1)$$

One has in Brownian conditions:

$$\overline{(x-\overline{x})^2} = Dt \qquad (6.22\text{-}2)$$

where bars denote averages and D is a constant. Since, however, the events are not independent, one has a correlation function for the random variable Q: .

$$C(s) = \frac{1}{\overline{Q^2}} \int_0^{\infty} Q(t)\,Q(t+s)\,\mathrm{d}t \qquad (6.23\text{-}3)$$

where the correlation time L_t is:

$$L_t = \int_0^{\infty} C(s)\,\mathrm{d}s \qquad (6.22\text{-}4)$$

If the integral in eq.6.22-3 converges (Markovian models) one has: (i) the original variable $Q(t)$ is still Gaussian in the long run; (ii) for the cumulative variable $q(t)$ one has:

$$\overline{q^2(t)} \sim t \qquad \text{for} \qquad t \gg L_\mathrm{t} \tag{6.22-5a}$$

$$\overline{q^2(t)} \sim t^2 \qquad \text{for} \qquad t \ll L_\mathrm{t} \tag{6.22-5b}$$

It is possible to assume that the integral in eq.6.22-3 does not converge. Then, the limit process leading to the asymptotic behaviour of the process outlined above cannot be carried out. In such cases, there is an infinite correlation time; the behaviour of q^2 lies between proportionality to t and proportionality to t^2. It is possible to formally develop mathematical models for such processes that always (for short *and* long time ranges) have a behaviour of the cumulative variable q as:

$$\overline{q^2} \sim t^{2H} \tag{6.22-6}$$

with H a constant between 0.5 and 1. This can be done by postulating that the process is self-similar in all scale intervals (Mandelbrot, 1965). The above model has been applied by Mandelbrot and Wallis (1968, 1969) to explain the occurrence of floods: the cumulative variable q was identified with the hydrograph variable; empirically a long-range behaviour of floods of the type of eq.6.22-6 had been observed by Hurst (1951). This behaviour has therefore been called "Hurst-phenomenon" and H has been termed the "Hurst-parameter". The procedure, however, was criticized by Scheidegger (1970e) and Rodriguez-Iturbe et al. (1972) on the grounds that the assumption of self-similarity does not appear to be physically justified and amounts to an unwarranted extrapolation from limited observations to short (where a catastrophe occurs in the spectrum) and long time intervals. Furthermore, the "Hurst-phenomenon" can also be obtained from other models.

The problem of finding stochastic models that describe observed hydrological time series is still far from solved. A test of various theoretical distributions in the light of data has been reported by Spence (1973) who found that for rivers of the Canadian plains a two-parameter log-normal distribution fitted the data most accurately. Similarly, Matalas and Wallis (1973) found that a Pearson-type-3 distribution suited their data best. Torelli and Chow (1972) found that hydrological time series are generally not stationary, but that they can be stationarized by, for example, dividing each monthly value by the estimate of the corresponding monthly standard deviation. However, the analysis then does not lead to particularly important features from a practical point of view (Torelli, 1973).

6.23. Flash floods

If one wishes to investigate the statistics of floods, the whole hydrograph $Q(t)$ of a gauging station does not have to be considered. A stream hydrograph can be regarded as a sequence of nearly instantaneous "flood" peaks which are separated by quite long periods of low flow (Dirby, 1969). One therefore has "events" of different "magnitude" Q (a random variable) occurring at random times. The situation is very similar to that encountered in seismology.

The "magnitude" of a flood is generally not given simply by the flow Q, but by the "exceedance" over the "normal" flow. One would like to predict the likelihood of a certain flood in a given time interval.

The *maximum flood* that may be expected is deduced from the flood hydrograph theory, as indicated in Sect.6.21, using the maximum estimated rainfall as input into the response function.

For the prediction of the *frequency of floods* a statistical approach has to be taken, similar to that used in seismology.

The raw data are arranged as shown in Table IX; it gives the number of floods that exceed given levels Q_i (Q_i-floods) in any one year for a number of years. For the total number of years Y for which records are available, a frequency–magnitude curve can be constructed which represents the number N_i of floods exceeding the magnitude Q_i. The annual occurrence S_i of a Q_i-flood is therefore:

$$S_i = N_i/Y$$

TABLE IX

Type of data required for flood-recurrence analysis; floods of river Gail at Rattendorf, Austria, during the period 1890–1952

N	Year	$Q > 400$ m^3/sec value	Number of peaks exceeding a flow of m^3/sec:								Q maximum m^3/sec
			400	350	300	250	200	150	100	66	
1	1890							1	2	8	153
2	91								4	8	133
3	92									6	98
4	93							1	1	1	167
5	94									2	74
6	95							1	1	3	191
7	96						1	2	5	7	236
8	97								1	5	127
9	98							1	3	9	175
10	99									2	76
11	1900									3	109
12	01							1	3	7	151
13	02								1	5	109
14	03			1	1	1	1	2	2	6	395
15	04								3	5	122
16	05							1	4	6	167
17	06						1	1	1	4	202
18	07						1	2	3	6	202
19	08								1	4	122
20	09								1	5	115
21	10								4	7	136
22	11							1	3	4	175
23	12								1	6	109
24	13								1	2	129
25	14	412	1	1	1	1	1	2	3	4	412
26	15									4	97
27	16								1	9	136
28	17							2	4	6	167
29	18								2	5	143
30	19								2	7	115
31	20							1	3	7	175
32	21									1	82
33	22							1	3	5	175
34	23								2	8	146
35	24							1	3	6	164
36	25					1	1	1	2	4	251
37	26			1	1	1	2	2	3	8	368
38	27					1	1	3	5	9	254
39	28			1	1	1	2	3	3	9	378

TABLE IX (continued)

N	Year	Q >400 m^3/sec value	Number of peaks exceeding a flow of m^3/sec:								Q maximum m^3/sec
			400	350	300	250	200	150	100	66	
40	29									1	66
41	30					1	1	1	1	4	254
42	31						1	1	5	11	232
43	32								1	3	109
44	33								1	3	136
45	34					1	2	3	6	8	254
46	35	540	1	1	2	2	4	4	4	8	540
47	36							1	2	2	159
48	37					1	3	4	7	11	266
49	38									4	95
50	39						1	1	2	3	208
51	40			1	1	1	1	2	3	3	361
52	41						1	1	2	4	226
53	42							2	2	5	190
54	43									1	71
55	44									2	80
56	45					1	1	1	1	1	257
57	46						1	1	2	4	212
58	47								1	5	101
59	48									6	85
60	49									1	71
61	50									3	92
62	51	522	1	1	1	1	2	2	6	6	522
63	52								1	5	103
			3	7	8	14	29	54	127	317	

The recurrence time T_i is then:

$$T_i = \frac{Y}{N_i} \tag{6.23-1}$$

The probability P_i of a Q_i-flood occurring in any one year is obtained by calculating the probability that one *or more* floods in any one year are Q_i-floods. This was done by Chow (1964, pp.8–22ff) assuming independent trials, who obtained:

$$P_i = 1 - e^{-S_i} \tag{6.23-2}$$

This formula is based on a limit procedure; the distribution of flood values in one year is immaterial. It can be shown that the difference between P_i and S_i becomes small as these values become small. Hence, it is possible to take for the recurrence time:

$$T_i = \frac{1}{S_i} \approx \frac{1}{P_i} \tag{6.23-3}$$

when $T_E > 5$ years.

The return period is the average period of return. There is about a 65% chance (assuming independent trials) that an event will actually occur within its return period. The raw data are often represented as "annual maximum flood" values. These are the maximum flow rates Q_i that have been observed in subsequent years numbered 1, 2, ...i... at a given gauging station. This type of data represents an equally spaced time series. The probability P_i for finding Q_i in any one year is precisely P_i as given by eq.6.23-2, the corresponding recurrence times T_i can be calculated from such a series by eqs.6.23-1 and 6.23-2.

The floods of given Q may be produced by a Poisson process (Kirkby, 1969). However, for the further development of the theory this is of no significance.

From the raw data, one can obtain a flood-recurrence curve (Fig.38), which gives the relationship between the average recurrence interval T_i and flood value Q_i. This curve can be used for design purposes. Of course, of primary interest is an extrapolation of such a curve to a range where it has not been measured. This can only be done if certain statistical hypotheses are made about the data series. Assuming the daily flow readings to be distributed exponentially, of which the extreme values (the annual maxima) are observed in m samples of equal size s (one year), Gumbel (1941, 1943), using the theory of extreme values, showed that for $m \to \infty$, $s \to \infty$, the cumulative probability P that any of

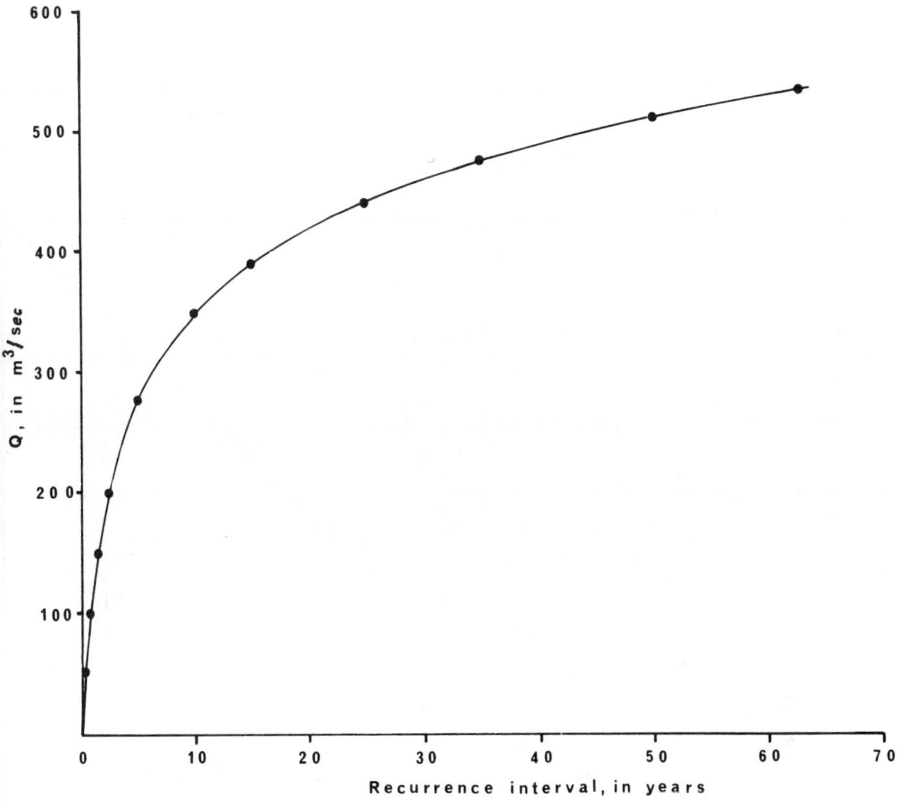

Fig. 38. Flood-recurrence curve for the river Gail, corresponding to the data of Table X.

the extremes will be less than Q is:

$$P(Q) = \exp\{-\exp[-a(Q)-u)]\} \tag{6.23-4}$$

where a and u are two parameters which can be determined from the available sample. The probability of the occurrence of a flood larger than Q in a year is thus $1 - P(Q)$; the recurrence time is therefore:

$$T_Q = \frac{1}{1-P(Q)} \tag{6.23-5}$$

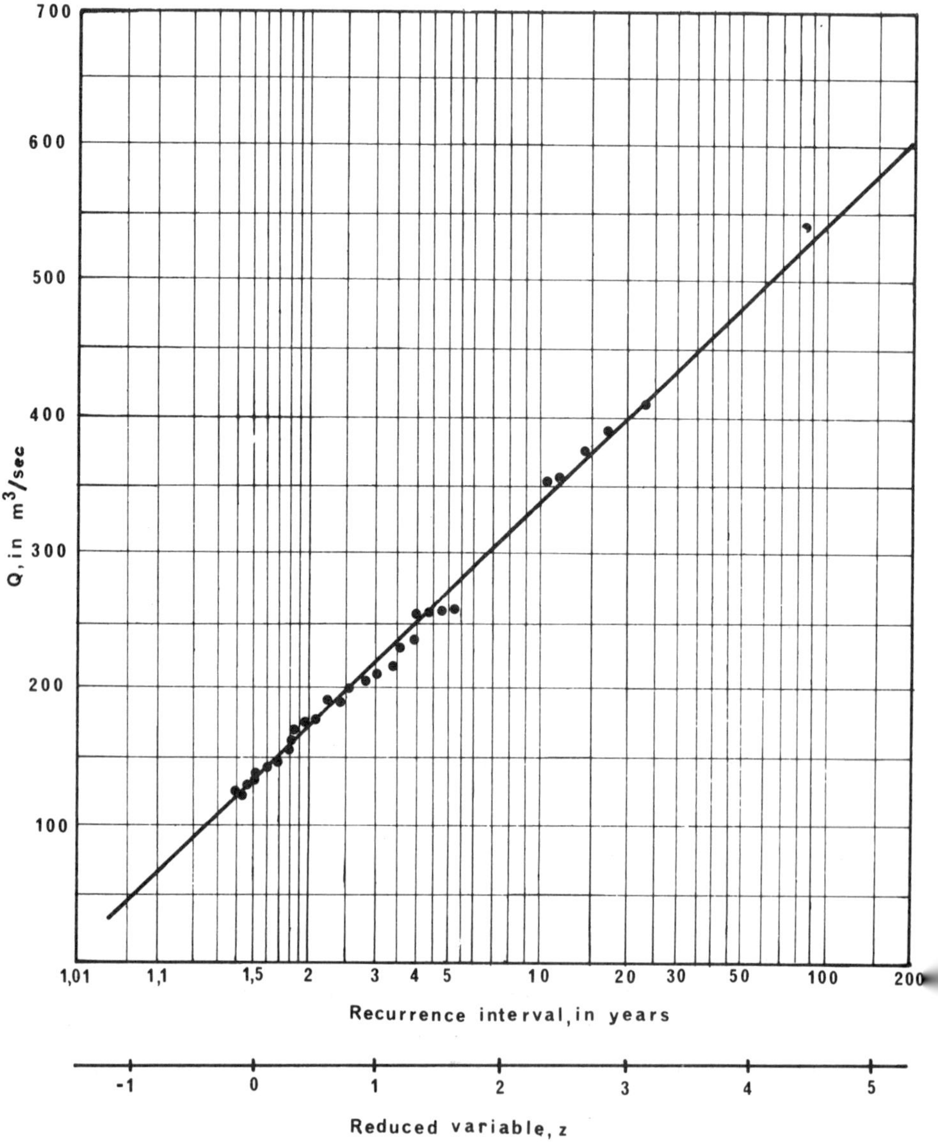

Fig. 39. Gumbel plot for the data of Table IX (River Gail, Austria).

or:

$$-a(Q - u) = \log\left[-\log\left(1 - \frac{1}{T_Q}\right)\right] \equiv z \qquad (6.23\text{-}6)$$

It is clear, then, that plotting Q against z should yield a straight line. For this purpose, "Gumbel paper" (Powell, 1943) has been created that is marked on the abscissa directly in T_Q. From the available data T_Q versus Q, the "Gumbel plot" can be drawn; it should be a straight line (Fig.39); the latter is extrapolated as noted previously, the maximum possible flood is calculated by the hydrograph method; its return period can be obtained from the Gumbel plot.

The Gumbel procedure is based on an asymptotic form of a problem of sampling theory (maxima in series of given lengths) of measurements of identically distributed variables, which may not be entirely justified. Todorović and his coworkers (Todorović, 1970; Todorović and Zelenhasić, 1970; Todorović and Rousselle, 1971; Todorović and Woolhiser, 1972) considered more general random samples of random numbers (nonidentically distributed variables). As a result they produced a refinement of Gumbel's theory.

6.24. Catastrophic flood waves

Of greatest geomorphological significance are catastrophic flood waves that proceed down a valley. These may be caused by the breaking of a dam (Malpasset, December 22, 1959) or when earth or rock slides into a lake (Vajont, October 9, 1963) and causes it to overflow. In all these cases, an almost instantaneous flood crest is created in the headwater region (a step function) of a valley; of interest is its subsequent behaviour in proceeding down a channel.

The problem is to calculate the behaviour of a nonstationary gravitational solitary water wave in a prescribed channel. The basic mathematical formulation of the problem was achieved long ago by De St. Venant (1871). The deduction of De St. Venant's equations is based on the assumption that the flow is one-dimensional with parallel stream lines. This is, of course, an oversimplification; it is justified if the curvature of the wave surface is small compared with the water depth and if the channel is long. The basic equations are then (e.g. Stoker, 1957):

$$\frac{\partial u}{\partial t} + u\frac{\partial u}{\partial x} + g\frac{\partial h}{\partial x} + g(S_e - S_s) + \frac{q}{F}(u - v\cos\beta) = 0 \qquad (6.24\text{-}1)$$

$$\frac{\partial F}{\partial t} + u\frac{\partial F}{\partial x} - F\frac{\partial u}{\partial x} - q = 0 \qquad (6.24\text{-}2)$$

Here u is the mean velocity at the time t in the cross-section at the point x along the valley, i.e:

$$u = Q/F \qquad (6.24\text{-}3)$$

where Q is the instantaneous flow rate and F is the instantaneous cross-section; furthermore, h is the water depth, S_e the slope of the energy line and S_s the slope of the bottom line, q is a constant lateral influx per unit length, v is the velocity of the laterally inflowing water and β the angle at which it enters.

The instantaneous cross-section F is a function of h and x; it must be given numerically from a planimetrization of the actual channel under consideration. The slope of the energy line can be calculated from any one of the usual river formulas.

Thus, the task is to integrate the system of equations (6.24-1/2) for given functions $F(h,x)$. This was impossible until the advent of computers. Generally, one uses numerical methods based on the method of characteristics. Algorithms for doing this have been reviewed by Lawler (1964), Liggett and Woolhiser (1967), Dracos (1970) and by Amein (1966, 1968) in discussions of the general problem of "flood routing". Specific cases, e.g., the flood wave in the Malpasset disaster, were calculated by Chervet and Dallèves (1970; Chervet, 1971) who found a reasonable agreement between the calculated and the observed values. The lateral influx of water may not only be due to tributary valleys, but also to bank seepage (Pinder and Sauer, 1971; Zitta and Wiggert, 1971).

6.25. Lakes

The significance of lakes in hydrogeomorphology lies first of all in the fact that they act as storage basins for flood crests proceeding downstream. Secondly, the water in a lake basin may be subject to internal oscillations (seiches) and finally, waves, currents and circulations may develop; these are not specific to lakes but are equally prevalent in seas and oceans. In this section, we shall deal with the first two aspects of lake hydrodynamics; the third aspect will be included in the section on shorelines and coasts.

Thus, the fluctuation of lake-level as a function of variable inflow is of interest. Let us denote the inflow by Q_i, the outflow by Q_e, then the continuity equation leads to (e.g. Hutchinson, 1957):

$$Q_i - Q_e = A \frac{dh}{dt} \tag{6.25-1}$$

where h is the height of water-level above a suitable base line, A the surface area of the lake and t time. The influx Q_i contains, in principle, surface, groundwater and precipitation influx; similarly, the outflow Q_e contains the efflux due to surface flow, groundwater flow and evaporation. However, for flood cases, the surface flow is the preponderant quantity; the other types of flow may be disregarded.

The influx Q_i is, thus, a given or (for an *expected* flood) assumed quantity. The outflow Q_e is a known function of h for a known lake basin; the same is true for the free surface A which is also a known function of h.

Therefore, it is seen that for the solution of the high-water retention problem of a lake eq.6.25-1 has to be solved. This equation is a total differential equation with empirically given functions $A(h)$ and $Q_e(h)$; the solution is therefore obtained by numerical or graphical methods. In finite differences, eq.6.25-1 can be written:

$$\frac{\Delta h}{Q_i - Q_e} = \frac{\Delta t}{A} \tag{6.25-2}$$

It is then a simple matter to calculate Δh for a given Δt starting at t_0, h_0; $h_0 + \Delta h$ is the new level at time $t_0 + \Delta t$ etc. Greater accuracy can be obtained by using Q_i at $t_0 + \Delta t/2$ rather than at t_0.

Some of the practical schemes have been discussed by Vischer (1970, 1971), Vischer and Moser (1971) and Tschopp (1972).

When considering the internal oscillations (seiches), we note that any basin has a series of eigen-frequencies for surface oscillations. These are gravity waves. In shallow water, the wave velocity c of such waves is given by:

$$c = \sqrt{gh} \qquad (6.25\text{-}3)$$

where g is the gravity acceleration and h the water depth. In a basin of length (or width) L, the wave length λ of a seiche must be $2L/n$ (with n integer) so that the beginning and end of the length L can be on a wave centre. From the general relation for waves one has:

$$\lambda/T = c \qquad (6.25\text{-}4)$$

where T is the wave period; one thus obtains the relation:

$$T_n = \frac{1}{n}\frac{2L}{\sqrt{gh}} \qquad (6.25\text{-}5)$$

for the period of the seiche with n nodes along the length L of the basin.

Refinements can be made in the above formulas by allowing for a non-rectangular basin and for a varied bottom-profile. To do this, one has to apply the hydrodynamic theory of waves. Since this is a basically nonlinear theory, with external conditions (basin geometry) given heuristically, various approximation and numerical procedures have been devised to treat the problem. Many of the attempts have been reviewed by Defant (1957). Applications of the theory to many lakes have been reported, e.g., by Caloi (1972; Caloi and Spadea, 1958, 1959) to the lakes of upper Italy; in these reports the theoretically predicted and actually observed periods generally agree well. The agreement improves with the complexity of the calculation procedure that has been employed.

Also of particular interest are seiches in semi-enclosed basins where there may be a coupling to the outside; this may be the case in harbours. Examples were analysed by Easton (1972; Sidney harbour, N.S., Canada) and by Donn and Wolf (1972; Grindavík harbour, Iceland).

The seiches in a lake or harbour are ordinarily minute events. However, if the exciting oscillations (edgewaves generated by storms, barometric pressure fluctuations) happen to resonate with one of the natural periods, dramatic and frequently destructive events may result: of this kind was the surge of Lake Michigan at Chicago on June 26, 1954 (Zumberge and Ayers, 1964). The seiches in Sidney, N.S. and Capetown, South Africa harbours have been troublesome enough (damage to ships) to have warranted engineering attention.

6.26. River-flood erosion

The geomorphological effects of the hydrological phenomena discussed so far in this chapter are caused by the fact that they are *deviations* from the equilibrium conditions. The geomorphological action of rivers in general can be split into bottom (channel) erosion and lateral erosion. The steady-state cases were summarized by the writer (Scheidegger, 1970a). We shall keep the same division here.

Regarding bottom (channel) erosion we recall that, in the steady state, rivers cause a transportation of sediment along their bed. Over the course of the years, the river is at grade, ripples alternate with pools, and differential transportation causes a regular gradation of pebbles along the profile. The various theories of "river bed processes" have been reviewed at length by the writer (Scheidegger, 1970a). The problem of riffles and dunes has been investigated more recently by Callander (1969) and by Engelund (1970) who demonstrated that the instability of the turbulent flow can well explain these phenomena. Similarly a study of the threshold of bed-load transportation has been made by Muller et al. (1971). Typical small-scale sedimentary structures have been studied, particularly by Allen (1968) who wrote a book on current ripples. Other studies on the same subject have been reported by Kennedy (1969). Allen (1971b) has also investigated the provenance of longitudinal bed marks.

The mechanical effect of transverse obstacles upon the eroding action of a river was investigated experimentally by Volkart (1972). If an obstacle exists in a river which reaches from beneath the river to the (original) river bed, a depression develops behind it whose greatest depth h (in m) can be estimated from the formula:

$$h = \frac{1.25\, q^{1/2}\, S^{1/2}}{d^{5/12}}\, L_t^{2/3}\, \frac{\gamma_w}{\gamma_s g^{1/4}} \,. \qquad (6.26\text{-}1)$$

Here, q is the specific water flux (total flow divided by channel width) in $m^2 sec^{-1}$, S the bed slope (number), d the characteristic bed particle size (m), L_t the distance between subsequent obstacles (m), γ_w the specific weight of water, γ_s the specific weight of the bed load under buoyancy and g the gravity acceleration (m sec^{-2}). The above formula is valid if no bed-load transport exists; h is smaller if bed-load is supplied from above. Up to a certain combination of the parameters, the depression remains stable. Beyond this point, however, a washout in front of each obstacle occurs (the deepest point wanders downstream to the next obstacle) which is then laid bare. The condition for stability is (Volkart, 1972):

$$\frac{q^{1/2} S^{1/2}}{d^{5/12} L_t^{1/3}} \leqslant 0.16\, g^{1/4}\, \frac{\gamma_s}{\gamma_w} \qquad (6.26\text{-}2)$$

The above formulas can be used for design purposes when works to stabilize a river are planned.

The nature of a geomorphologically significant hydrological event represents a *deviation* from the steady state: generally the event is connected with an increased flow. The general effect of such an increased flow is clear. For a short period of time, the carrying capacity of the river is greatly increased. This manifests itself in the increased erosion of large material in the headwater region, and in the deposition of this large material further downstream. In other words, the pebble-gradation curve is shifted downstream. The river profile is also affected ephemerally because of the general relation between bed-load transport and channel shape (Wilcock, 1971).

The above statements are rather qualitative. Nevertheless, the general character of the predictions has been verified by Bradley et al. (1972) in an investigation of the effects produced by the flood flows of the Knik River in Alaska. Coarse boulders have been found that were carried a great distance. However, accurate investigations of the theoretical-mechanical aspects of the geomorphic devastations to be expected by river floods do not appear to have been carried out yet, with the exception of Bradley et al. (1972) who have made some scaling experiments in an attempt to explain their observations.

What has been said above about flood channel erosion, also applies, *mutatis mutandis*, to lateral (bank) erosion. The equilibrium configuration of a river at grade is a sequence of meanders (see e.g. Wundt, 1949; Langbein and Leopold, 1966; Thakur and Scheidegger, 1970). It is reasonable to assume that floods will be the main agents of change in an established meander pattern: they cause break-throughs which form oxbow lakes, and widen loops, etc.

The writer is not aware of any theoretical or experimental studies of the mechanics of such violent changes. The available investigations deal with equilibrium conditions. Only Schumm and Khan (1972) made an experimental study of the effect of changes, albeit slow ones, in the external conditions (bed-load, climate, etc.) but not of catastrophic changes, such as would be represented by the passage of a flood wave through and across the meanders. The subject is therefore still wide open.

6.27. Water erosion on slopes

A heavy rainfall or snow-melt may not only cause floods in a river channel, but may also have a direct action upon slopes.

A slope may be considered simply as an impermeable (or partly permeable) inclined plane upon which a sheet of water develops which, then, flows downhill as a "sheet flood". This sheet flood, in turn, will act upon the underlying material. The dynamics of the build-up, and the decay of such a sheet flood, due to a steady rainfall of finite duration, was studied by Henderson and Wooding (1964). Accordingly, the dynamics of the water sheet is determined by a continuity equation and an equation of motion. The continuity equation is:

$$\partial q/\partial x + \partial h/\partial t = v_0 \qquad (6.27\text{-}1)$$

where q is the flow or discharge rate, h the depth of the water, v_0 the rainfall rate per unit area, x the down-slope coordinate and t time. Seepage into the plane could also be taken into account. The equation of motion is:

$$q = \alpha h^{\mathrm{n}} \qquad (6.27\text{-}2)$$

where there are two different cases:

(a) Laminar flow:

$$\alpha = gS/(3\nu)\,, \qquad n = 3 \qquad (6.27\text{-}3a)$$

where g is the gravity acceleration, ν the kinematic viscosity and S the slope gradient;

(b) Turbulent flow:

$$\alpha = C_1 S^{1/2}\,, \qquad n = 3/2 \qquad (6.27\text{-}3b)$$

where C_1 is the Chézy coefficient.

If eq.6.27-2 is inserted into eq.6.27-1, a partial differential equation for h (hyperbolic type) is obtained which has to be solved for the appropriate initial and boundary conditions; this is most conveniently done by the method of characteristics. During the build-up phase it appears that a plateau parallel to the slope rises with a constant velocity v_0, until a steady state is reached which is given by the curve:

$$\alpha h^{\mathrm{n}} = v_0 x \qquad (6.27\text{-}4)$$

The situation is illustrated in Fig.40, in which AB, AD_1E_1, AD_2E_2, ACB represent subsequent profiles. Referring to Fig.40, the steady

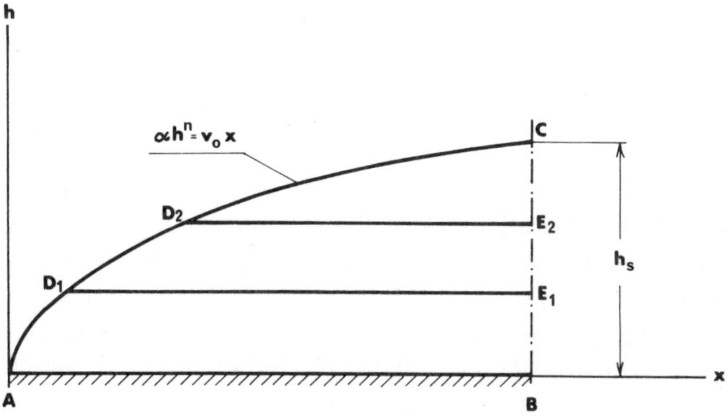

Fig. 40. Profile during the build-up phase of a sheet flood. (After Henderson and Wooding, 1964.)

state is reached after a time t_s:

$$t_s = h_s/v_0 = L/\alpha h_s^{n-1} \tag{6.27-5}$$

where h_s is the depth at point B. The outflow at B is:

$$q = \alpha h^n = \begin{cases} \alpha v_0^n t^n & (t < t_s) \\ L v_0 & (t > t_s) \end{cases} \tag{6.27-6}$$

The decay phase (after cessation of the rain) can also be calculated, but this is more complicated.

The erosional activity of a sheet flood can be calculated according to one of the drag force equations for rivers (cf. Scheidegger, 1970a). The forces of flowing water can be substantial, as evidenced by the devastations caused by catastrophic floods in plains. These floods resemble most closely the picture of a "sheet" flood considered above. Nevertheless, the damage caused by such floods is generally due, to a large extent, to the fact that the water enters cellars and houses, not to the velocity of the flood. Thus, the water depth h is of more significance than the velocity v.

With the exception of floods in plains, the cases encountered in nature are generally quite different from the theoretical picture of a "sheet flood", inasmuch as innumerable gulleys are immediately formed. This is most readily visible in the erosional features of badlands or heaps of mine tailings. Gerber and Scheidegger (1973) have analysed the characteristic features of such erosional gulleys, and have shown these gulleys are recognized by the fact that they have the features of bifurcating topological trees. The actual *collapse* of the features created by the gradual erosion of the gulleys (caused by the usual drag-forces of the flowing water), i.e., of the "relics" *between* the gulleys, occurs because the statically stable stresses are eventually exceeded (Moser, 1971); cases of such, sometimes very dangerous, collapses that occurred in a sand quarry have been described in the cited paper of Gerber and Scheidegger.

6.28. Tectonic effects of surface-water

Surface-water can have tectonic effects which may be quite noticeable. The filling of artificially created lakes may induce considerable additional stresses in the earth's crust which may have diastrous consequences. These stresses and the corresponding displacements have been calculated by Lee (1972) for a layered elastic half-space upon which a non-homogeneous load (corresponding to a lake) is applied from above. Simpler models had been calculated earlier by Steinhauser (1934) and Caputo (1961). It was found that the additional stresses, which are induced, e.g., by a lake of $1.5 \cdot 10^{10}$ tons water weight, are noticeable as far down as the Mohorovičić discontinuity. For a rigidity of $6.3 \cdot 10^{10}$ dynes/cm^2 and a Poisson ratio of 0.25, the maximum displacement at the surface is about 7 cm for a lake as described above.

The stresses in the earth's crust which are induced by water loads (artificial lakes), may trigger earthquakes. A survey of cases where this occurred has been given by Nikolayev (1972) and by Caloi (1970). In particular, the rapid fluctuations of the water level cause earthquakes (Båth, 1970), mostly during the loading phase (Rothé, 1969b, 1970). In time, however, correlations are also observed with the unloading phase (Carder, 1970). Instances where this occurred have been observed in the United States (Hoover Dam: Carder, 1970), Spain (Canelos Dam), Greece (Cremasta Lake: Comniakis et al., 1968), India (Koyna Dam: Gubin, 1970; Aggarwal, 1972; Chopra and Chakrabarti, 1973),

Pakistan (Mangla Dam: Adams, 1968, 1972), South Africa (Kariba Lake: Gough and Gough, 1970) and other countries. Sometimes the induced earthquakes registered nearly 6 units on Richter's magnitude scale, and caused loss of human life and severe damage. Seismic activity is particularly likely to be induced if the water depth of the reservoir is 100 m or more. Not all reservoir fill-ups cause earthquakes; apparently the geotectonic conditions have to be favourable: a triggering effect releases the stresses only if they have been already accumulated.

The above instances refer to the creation of artificial lakes. However, surface water can cause earthquake swarms under natural conditions too. It is again the load due to the weight of the water which causes these phenomena.

For example, since 1900, numerous earthquakes occur in the Mississippi valley within a range of 200 miles of new Madrid (Mo.) which have been felt with an intensity of up to 7 units on the Mercalli scale, but have been experienced with 12 intensity scale units (total destruction) at New Madrid in 1811 and 1812. A careful study of the correlation between these earthquakes and the level of the Mississippi river has been made by McGinnis (1964), who plotted the mean monthly levels against the observed bimonthly frequency of earthquakes. The result of this study showed that there is a good correlation between earthquake frequency and the *derivative* of the river level. The phenomenon can be explained by the changing load of the river. The difference in height between the highest and lowest river levels is a little more than 10 m, resulting in stress changes of up to $5 \cdot 10^5$ dynes/cm^2 at depth. It is conceivable that such stress changes will trigger earthquakes.

In this connection, Lursmanashvili (1973) suggested that the correlation between solar activity and earthquake activity mentioned in Sect.2.22 is, in effect, hydrological in nature: the varying solar activity could cause variations in the static loads upon certain regions of the earth due to various amounts of ice and water being present at different times (notably in the Caucasus area where the water-level in the nearby Caspian Sea was found to be subject to secular fluctuations). The varying static loads, then, would trigger the earthquakes.

6.3. Geomorphic effects of ground-water

6.31. Principles of seepage

Ground-water is the water which does not flow on the surface of the earth, but, rather, flows within it. Like surface-water, it can have various geomorphological effects.

The ground-water occupies the pore space of the soil and rocks, forming an aquifer. The aquifer may be confined or unconfined. A confined aquifer is bounded above and below by impermeable layers; an unconfined aquifer is bounded only at the bottom by an impermeable layer; above, the ground-water forms a *free surface*.

The flow of ground-water is seepage flow, governed by Darcy's law (see Scheidegger, 1974). In a differential form this law can be written as follows:

$$\boldsymbol{q} = -\frac{k}{\eta}(\text{grad}\, p + \rho \boldsymbol{g}) \tag{6.31-1}$$

where $\boldsymbol{q}$ is the seepage velocity vector (given as the volume seeping in unit time through a unit area), k the permeability (a constant characterizing the porous medium), η the viscosity of the percolating water, p the pressure in the liquid in the pores, ρ the density of the liquid and $\boldsymbol{g}$ the gravity acceleration vector.

Darcy's law alone is not sufficient to determine a flow problem; in addition, a continuity equation is needed:

$$-P\frac{\partial \rho}{\partial t} = \text{div}\, \rho \boldsymbol{q} \tag{6.31-2}$$

where P is the porosity (a fractional pure number characterizing the porous medium; it is equal to the average ratio of pore volume to bulk volume) and t time. Combined, eqs.6.31-1 and 6.31-2 yield:

$$P\frac{\partial \rho}{\partial t} = \text{div}\left[\rho \frac{k}{\eta}(\text{grad}\, p - \rho \boldsymbol{g})\right] \tag{6.31-3}$$

which is a partial differential equation for ρ. If the connection between ρ and p (the rheological equation of the fluid):

$$\rho = \rho(p) \tag{6.31-4}$$

is added, a flow problem is determined, provided the relevant initial and boundary conditions are known.

Any ground-water flow problem, at least in a first approximation, is described by a solution of the flow equations presented above. Under quite general conditions, eq.6.31-3 can be approximated by a diffusivity equation, so that many ground-water flow problems can be solved by finding the appropriate solutions of a diffusivity equation.

The description of ground-water flow by Darcy's law may sometimes need to be further refined. The capillary pressure (the "suction" in the undersaturated zone in soil physics), nonlinear deviations from the linear flow law (in flow regions corresponding to a Reynolds number of more than 1) and molecular effects may have to be taken into account. For geomorphological applications this is, however, not very important.

6.32. Effect of pore-water on slope stability

One of the main effects of pore-water on a material is the collapse of its mechanical resistance. In a granular material, the yield conditions are given by the Coulomb equation (cf. 4.15-1):

$$\tau_s = \sigma \tan \phi + c \tag{6.32-1}$$

where τ_s is the shearing limit, σ the normal ("overburden") stress, c a constant indicative of the cohesion of the material and ϕ another material constant, called the angle of internal friction. In a water-saturated assemblage of grains, the basic relation 6.32-1 is changed. The relation for the shearing resistance is (after Terzaghi and Peck, 1948; see also Bishop, 1959, and Bishop and Bjerrum, 1961):

$$\tau_s = c + (\sigma - p_w) \tan \phi \tag{6.32-2}$$

Thus, all one has to do is to replace the normal stresses by the effective stress σ_e, given by:

$$\sigma_e = \sigma - p_w \tag{6.32-3}$$

in the usual equations for slope stability. It is seen that the shearing resistance may become zero if p_w becomes large enough; in fact, eq.6.32-2 breaks down for negative σ_e. Under such conditions, the porous medium undergoes "splitting" failure. Thus, slopes that are stable under dry conditions may become unstable when water-logged: it is a well-known fact that protracted periods of rain (by raising the pore-pressure) may trigger landslides. A similar rise in pore-pressure caused by the filling up of the artificial lake in the Vajont valley (Italy) may have been a contributing cause of the catastrophic slide that occurred there on 9 October, 1963. A similar, but not quite as catastrophic, case near Gepatsch in Austria was described by Lauffer et al. (1971); large movements on a slope were evidently triggered by the pore-pressure induced by the filling up of an artificial reservoir in this case.

The situation in rock is similar to that in soild masses. The "effective pressure law" (eq.6.32-3) has been tested repeatedly and has been found to apply to crystalline rocks of low porosity (Brace and Martin, 1968) as well as for soil. However, the joint system in rocks may have to be treated separately. A review of this case has been given by Serafim (1969). The flow in the joints has to be taken as a separate flow-system, and the interaction between the flow in the joints and the flow in the pores has to be determined. Needless to say, this requires a highly sophisticated mathematical apparatus.

6.33. Consolidation

The phenomenon of consolidation is another geomorphological effect of ground-(pore)-water. Consolidation represents a two-fold process: the (porous) medium is undergoing finite deformations whilst the water contained in it is being displaced. A complete theory of the deformation of porous media during flow in the pores has been provided by Biot (1941, 1955). However, in consolidation studies the flow of the fluid *per se* is only of minor interest. Thus the flow velocity is eliminated altogether and the remaining equations describe the deformation of the mass as a whole. The standard way to do this (Terzaghi, 1943) is to assume that the effective pressure (cf. eq.6.32-3) which, under constant overburden pressure, is a linear function of the pore-pressure, affects the porosity *linearly*. The pore-water movement is then calculated according to Darcy's law which leads, as noted in Sect.3.1, under quite widely applicable conditions (Scheidegger, 1974) to a diffusivity equation. Because of the assumed *linear* connection between pore-pressure and porosity, one obtains a diffusivity equation for the description of the space–time dependence of the consolidation process under an imposed load.

In detail, the consolidation theory can be set up as follows. Let us consider a horizontal layer of thickness T of a consolidation-prone material. We assume with Terzaghi that the change of the ratio ϵ of void to grain volume is a linear function of the effective pressure p_e:

$$\frac{d\epsilon}{dp_e} = a \tag{6.33-1}$$

Since:

$$p_e = p_{\text{overburden}} - p_{\text{liquid}} \tag{6.33-2}$$

one has equivalently:

$$-\frac{d\epsilon}{dp_{\text{liquid}}} = a = \frac{d\epsilon}{\partial p_e} \tag{6.33-3}$$

or:

$$\frac{\partial \epsilon}{\partial t} = -a\,\frac{\partial p_{\text{liquid}}}{\partial t} = a\,\frac{\partial p_e}{\partial t} \tag{6.33-4}$$

The ratio ϵ, commonly used in consolidation theory, is related to the otherwise more customary porosity P by the following well-known obvious relation:

$$\epsilon = \frac{P}{1-P} \tag{6.33-5}$$

A change of porosity causes a reduction of the thickness of a thin slice which can be calculated as follows. The bulk volume of the porous medium is the sum of the pore- and grain-volumes:

$$V_{\text{bulk}} = V_{\text{grain}} + V_{\text{pore}} = (1-P)\,V_{\text{bulk}} + PV_{\text{bulk}} \tag{6.33-6}$$

Thus, if T be the thickness and A the area of the slice:

$$TA = V_{\text{bulk}} = \frac{V_{\text{grain}}}{1-P} \tag{6.33-7}$$

Assuming that the volume and the area remain constant during consolidation, we have:

$$\frac{\mathrm{d}T}{\mathrm{d}t}A = \frac{\mathrm{d}}{\mathrm{d}t}\,\frac{V_{\text{grain}}}{1-P} = -V_{\text{grain}}\,\frac{\partial P/\partial t}{(1-P)^2}$$

$$= -\frac{V_{\text{grain}}}{1-P}\,\frac{1}{1-P}\,\frac{\partial P}{\partial t} = -V_{\text{bulk}}\,\frac{1}{1-P}\,\frac{\partial P}{\partial t} \tag{6.33-8}$$

or, since $V_{\text{bulk}} = TA$:

$$\frac{1}{T}\,\frac{\partial T}{\partial t} = -\frac{1}{1-P}\,\frac{\partial P}{\partial t} \tag{6.33-9}$$

We note that (from eq.6.33-5):

$$\frac{\partial P}{\partial t} = (1-P)^2\,\frac{\partial \epsilon}{\partial t} \tag{6.33-10}$$

so that eq.6.33-9 can be written as:

$$\frac{1}{T}\,\frac{\partial T}{\partial t} = m\,\frac{\partial p_{\text{liquid}}}{\partial t} = -m\,\frac{\partial p_{\text{e}}}{\partial t} \tag{6.33-11}$$

with:

$$m = a(1-P) = \frac{a}{1-\epsilon} \tag{6.33-12}$$

This quantity is called the coefficient of consolidation.

We now consider the following problem. At $t = 0$ the overburden pressure has been changed by the additional amount p_1 from an original equilibrium pressure p_0.

This causes a hydrostatic super-pressure p_n in the pore fluid which, during the consolidation, will slowly disappear. At the beginning, it must be equal to p_1. Because of eq.6.33-2 we have:

$$\frac{\partial p_e}{\partial t} = -\frac{\partial p_n}{\partial t} \tag{6.33-13}$$

such that, with eq.6.33-11:

$$\frac{1}{T}\frac{\partial T}{\partial t} = m\frac{\partial p_n}{\partial t} \tag{6.33-14}$$

The continuity condition, then, requires that this consolidation causes a flow of fluid upward with seepage velocity q so that:

$$\frac{\partial q}{\partial z}\,dz = \frac{1}{T}\frac{\partial T}{\partial z}\,dz = -m\frac{\partial p_n}{\partial t}\,dz \tag{6.33-15}$$

Furthermore, Darcy's law requires:

$$q = \frac{k}{\rho g \eta}\frac{\partial p_n}{\partial z} \tag{6.33-16}$$

Combining the last two equations yields:

$$\frac{k}{\rho g \eta m}\frac{\partial^2 p_n}{\partial z^2} = \frac{\partial p_n}{\partial t} \tag{6.33-17}$$

which presents the fundamental equation of consolidation.

The various solutions for the appropriate boundary conditions will yield the characteristics of the consolidation processes. Thus, Terzaghi and Fröhlich (1936) gave a solution for freely draining and completely impervious boundaries, Bishop and Gibson (1964) for symmetrical impeded drainage conditions, Gray (1945), Huang (1969) and Schiffman and Stein (1970) for multi-layered systems and Mesri (1973) for partly impeded drainage.

These solutions can be interpreted as macroscopic creep equations: under the additional pressure p_{no} the change of the thickness T of a layer due to consolidation has the time-dependence as calculated by Terzaghi and Fröhlich (1936).

Accordingly, if a load is suddenly imposed at $t = 0$ upon a consolidation-prone material, and if one denotes the ratio of instantaneous settlement $s(t)$ and final (complete) settlement $s_1 = s(\infty)$ by ψ:

$$\psi = \frac{s(t)}{s_1} \tag{6.33-18}$$

one has:

$$\psi = \text{const}\,\sqrt{t} \tag{6.33-19}$$

for $\psi < 0.526$, and:

$$\psi = 1 - e^{-\text{const}.t} \tag{6.33-20}$$

for $\psi > 0.526$. It is seen, thus, that the response of a settling mass is asymptotically that of a Kelvin body.

The consolidation theory as discussed above assumed that there is a *linear* relationship between pore-pressure and porosity. More complicated models have been studied by Bjerrum (1967) and by Garlanger (1972) who investigated materials which exhibit creep under constant effective stress, and by Parlange (1973) who considered non-Darcian flow of the fluid.

A specialized form of consolidation occurs if the process can be considered as very slow, i.e., as quasi-stationary. It may then be termed *compaction*, particularly if only the end-product of the process (consolidated sediments) is of interest. Upon this basis, Marsal and Philipp (1959) set up a model according to which the compaction process of sediments occurs in three stages. In the first stage, fresh material is deposited in water leading to a loose assemblage of grains with high porosities (35–80%). In the second stage, the overburden builds up and the material begins to compact. In the third stage, the reduction of the porosity is caused by the *crushing* of the individual grains.

All these stages may be represented heuristically by an empirical expression of compaction defined by:

$$dP/d\sigma = -\kappa P \qquad (6.33\text{-}21)$$

where σ is the overburden pressure and P the porosity. This empirical expression shows κ as a function of P and σ. A well-known relation of this type is:

$$\kappa = \frac{\ln(P_0/P)}{\sigma - 1} \qquad (6.33\text{-}22)$$

If this relation is inserted in eq.6.33-2 and the corresponding integration is performed, ony obtains porosity–depth curves for consolidated sediments. It also explains the subsidence observed if fluid is withdrawn from underground. Numerous cases of this type have been collected by Poland and Davis (1969). Similar cases were also collected by Prokopovich (1969, 1971) who was also able to make several quantitative statements about the course of the subsidence. Naturally, the rate of subsidence depends on the withdrawal rate of underground fluids. However, once the withdrawal has stopped, there is a time lag before the ultimate subsidence is reached. For this time lag, Prokopovich used an exponential decay law of the following type:

$$Z - y = A \exp(-Bt) \qquad (6.33\text{-}23)$$

where y is the subsidence after the date when the withdrawal has stopped, t is time, and Z the ultimate subsidence. A and B are constants. From a plot of the observed data and a fit of the curve corresponding to eq.6.33-23, the constants A, B and the related ultimate subsidence Z can be calculated.

A particularly notorious case of land subsidence is occurring in Venice, Italy, that may, in part, be caused by the withdrawal of ground-water for industrial purposes. A study of this case has been made by Gambolati (1972a) based on a one-dimensional Terzaghi-type model; it was later extended (Gambolati, 1972b; Gambolati and Freeze, 1973) to three-dimensional Biot-type models. The data for Venice, however, are not very unequivocal and, in fact, the "subsidence" may to a large extent be caused by a rise in sea-level (cf. Sect.6.42) as well as by the withdrawal of groundwater.

An interesting occurrence is the inverse case to that discussed above: consolidation caused by the *wetting* of dry material rather than by the withdrawal of fluids from the ground (Lofgren, 1969). This "hydrocompaction" is presumably due to the lubrication of, and the reduction in, intergranular strength in loose grains caused by the addition of water. A proper mechanical model of the process, however, does not yet appear to exist.

6.34. Sink-holes

A special type of "settling" occurs in limestone terrains. Because of the leaching effect of water, large underground caverns can arise which, at some point, may lose their stability. The result is an underground collapse which may reach to the surface of the earth, forming a "sink-hole". Sink-holes may assume catastrophic proportions, as was experienced in several places in Alabama (Powell and LaMoreaux, 1969; Newton and Hyde, 1971); and in India (Srikantiya and Bhargava, 1972); in one instance in Alabama a sink-hole 1,275 m long, 1,050 m wide and 450 m deep was observed. Needless to say, a sudden collapse of this size represents a major accident in a built-up area.

It may be thought that the subsidence of an underground cavity will make itself felt on the surface along shear planes having the angle α with the horizontal (cf. Fig.41).

The angle α of the shear planes can be calculated as follows: the stress σ_n normal to the subsidence plane is (if σ_x, σ_z represent horizontal and vertical principal stresses, respectively) given by (cf. Sect.4.15):

$$\sigma_n = \sigma_z \sin^2\alpha + \sigma_x \cos^2\alpha \qquad (6.34\text{-}1)$$

The effective shear S along the shear plane is equal to the "static" shear (due to σ_z, σ_x) less the "frictional" shear (due to the internal friction of the material, which is described by the frictional

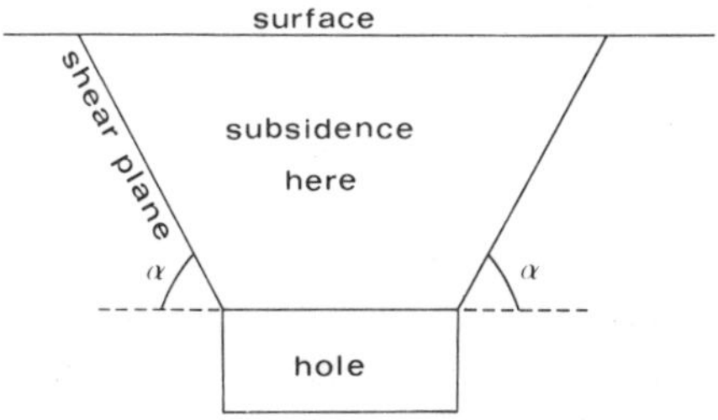

Fig. 41. Shear planes above a collapsing cavity, leading to a "sink-hole".

angle φ); thus:

$$S = \tfrac{1}{2}(\sigma_x - \sigma_z) \sin 2\alpha - \tan \varphi \cdot \sigma_n$$

$$= \tfrac{1}{2}(\sigma_x - \sigma_z) \sin 2\alpha - \tan \varphi \, (\sigma_z \sin^2\alpha + \sigma_x \cos^2\alpha) \qquad (6.34\text{-}2)$$

The maximum effective shear is reached when $dS/d\alpha = 0$. The angle α will be that angle for which S is maximum; therefore α can be calculated as follows:

$$\tan 2\alpha = -\frac{1}{\tan \varphi} = -\cot \varphi = \tan(\varphi + 90°) \qquad (6.34\text{-}3)$$

or:

$$\alpha = \tfrac{1}{2}\varphi + 45° \qquad (6.34\text{-}4)$$

The sink-hole will be formed, therefore, with a shear-plane angle α as calculated above.

6.35. Geodynamic effects of pore-water

The basic failure criterion for a granular material containing pore-water (eq.6.32-2) leads to the assumption that there is a relation between pore-water pressure and geodynamic phenomena: if a rise in pore-water pressure can reduce the shearing resistance of the material to zero, catastrophic failures may result.

Indeed, changes in pore-pressure (through the injection of wastes) seem to have been responsible for an earthquake swarm near Denver, Colorado (Healy et al., 1968; Bardwell, 1970; Evans,

1970) and near Rangeley, Colorado (Raleigh et al., 1972; Witherspoon and Maini, 1973). However, no corresponding correlation between the *withdrawal* of underground fluids and seismic activity was found by Sylvester et al. (1970). The occurrence of some of the earthquakes near large reservoirs, which was ascribed to surface-water loading in Sect.6.28, may, in fact, also have been aided by the increasing pore-water pressure in the rocks beneath these reservoirs when they were filled (Kissin, 1971). Nur and Booker (1972) suggested that aftershocks of large shallow earthquakes may be conditioned by subsurface flow: the main earthquake would cause a drop of the pressure in the ground-water which would slowly be restored by seepage. When higher pressures are regained, aftershocks would occur.

Earthquakes are the expression of a faulting process. A major role of pore-pressure in faulting has been postulated by Hubbert and Rubey (1959; Rubey and Hubbert, 1959). According to these authors the thrust sheets and nappes in folded mountains are made mechanically possible by the pore-pressure when it reaches similar values as the overburden pressure. Direct observations of fault movement in relation to the local rainfall in Italy have been reported by Caloi and Migani (1972), who believed in a direct correlation. Similarly, a study of fault stability and pore-pressure has been made by Byerlee and Brace (1972) in an experimental model in which the formation was induced to fracture by varying the pore-pressure in actual rocks under realistic conditions.

6.36. Quick sand and quick clay

When the pore-pressure p_w becomes equal to the overburden pressure σ in a cohesionless ($c = 0$) material, an inspection of eq.6.32-2 shows that the shearing resistance τ_s becomes zero. This happens in "quick sand". A sand-bed may be solid at one time, quick at another time, depending on the instantaneous flow regime. A bed that becomes quick in this way can lead to veritable geomorphic catastrophes.

Related to quick sands are "quick clays". However, clays are generally quite impermeable, although the porosity and therewith the ground-water content may be considerable. It is not Darcy's law which governs the movement of water particles in clays, but rather the osmotic diffusion. Therefore, clay can become "quick" with the alteration of the ionic composition (changing the osmotic potential) of the interstitial water. This effect, represented by the leaching of salt from marine clays, seems to be the cause of many quick-clay slides in Norway (Rosenqvist, 1953; see also Sect.4.73 of this book).

6.4. Shorelines and coasts

6.41. Equilibrium condition

The geomorphological aspect of a shoreline or coastline is the result of a steady-state process occurring in the vicinity of the shoreline or coastline. Basically, it is the near-shore circulation system which determines its configuration; much of the pertinent information has recently been summarized by the writer (Scheidegger, 1970a). Accordingly, the basic physical principle of coastline formation lies in an equilibrium between the long-shore material transport and the beach. If this dynamic equilibrium is changed in any way, drastic changes may take place on a coast (Davis and Fox, 1972). The connection between the variables involved can be investigated by process-response studies.

The near-shore circulation occurs in the form of cells which include rip currents at their boundaries and are connected with the cuspate structure of a coast. Recent studies on this question have been reported by Bowen (1969) and Komar and Inman (1970; Komar 1971). The general structure of the near-shore circulation cells has been investigated theoretically and experimentally in wave basins. Net long-shore currents are created when the equilibrium balance in each cell is not complete (Longuet-Higgins, 1970). The use of suitable models with regard to frictional coefficients etc., shows a relation between the long-shore current and wave parameters. It is directly related to the variation of breaker height along the shore.

6.42. Sea-level changes

An alteration in the relative position of the sea-level with regard to the land can cause severe geomorphological effects. As indicated, it is the *relative* position of the sea-level which is important; a change can be brought about by an absolute change in sea-level height or by an absolute change in the level of the land. In this, it is particularly the case of a rise of the sea-level ("sinking" of the land) which has the most severe effects.

World-wide, there seems to have been an overall rise in sea-level during the past 17,000 years. This has been established by carbon-14 dating of organic materials whose provenance is from animals and plants which are known to have lived close to sea-level. Data on this problem have been collected by Shepard (1963); they were later extended by Milliman and Emery (1968) who found a lowering of sea-level from 35,000 to 17,000 years before present, so that a low point may have been reached about 17,000 years before present. This may have coincided with the greatest extent of glaciation during the last ice age when a great amount of water was bound on the land in the ice-shields which existed then. Fig.42 shows a mean curve drawn for the data collected by Milliman and Emery (1968).

For present-day studies, the course of the sea-level curve in approximately the last few thousand years is of importance. As is

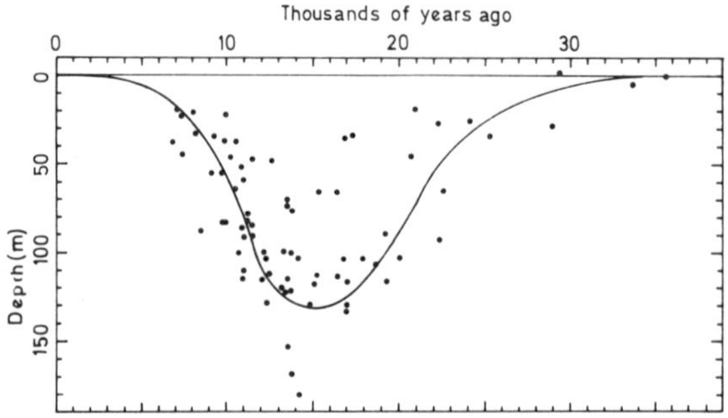

Fig. 42. Mean sea-level during the last 40,000 years. The dots represent individual measurements. (Drawn from data collected by Milliman and Emery, 1968.)

evident from the above, there has been a rise during that time; indications are that the rate of the latter has been 6 m in 6,000 years, i.e., 1 mm/year (Shepard, 1964). The above value is deduced for "stable" areas; it is a world-wide average that may not apply to specific areas. For example, it is well known that large parts of Scandinavia have risen substantially during the past 12,000 years or so, owing to an isostatic readjustment to the ice melting in this region after the end of the latest glaciation. Therefore, geodynamic effects in a particular locality may be the determining factor whether the sea advances or recedes at that place. The data on contemporary vertical movements on a world-wide scale have recently been collected by Gopwani and Scheidegger (1971) who have shown that these movements may also be (up or down) of the order of millimetres per year; thus the net effect between sea-level rise and vertical movements may be in millimetres per year, up or down.

Additional effects may be caused by local compaction and consolidation in conformity with the regime of pore fluids (cf. Sect.6.33). The settling of the ground in certain areas and a corresponding invasion of the sea in a particular locality may have to be ascribed to such causes.

As noted above, it is the *invasion* of the sea which is of most concern to mankind. A notorious case is the gradual flooding of the city of Venice in Italy (cf. Caputo et al., 1972) with catastrophic consequences. At that locality, the mean sea-level appears to be rising at just about the world rate of approximately 1 mm/year. Similar rates have been found at other coastal points of Italy. Hence, there is probably little one can do to stop the process of land "subsidence", inasmuch as it is the result of the general world-wide rise of sea-level due to the melting of the glaciers. In order to save the city, the lagoon would have to be closed off.

In other places, the subsidence may be man-made, mostly due to the withdrawal of underground fluids. Cases of this type have been discussed in Sect.6.33.

6.43. Large-scale changes on coasts

Observed large-scale changes occurring on coasts are essentially due to a rise or lowering of the sea-level. The changes caused by a rise (advance) in sea-level are of particular importance because they can cause severe damage.

On a beach, the changes caused by a rising sea-level are due to the displacement of the near-shore circulation system shoreward. Schwartz (1968) has described the mechanism. According to Schwartz, the material "eroded" from the upper reaches of the beach is deposited in the lower reaches and the near-shore zone in the water to restore the equilibrium that was present before the rise in sea-level. The equilibrium profile is thus simply displaced upward (corresponding to the rise in sea-level) and shoreward (to provide the material for the upward displacement of the near-shore zone).

On steep slopes, the mechanism of erosion is different, although here, too, the development of a new equilibrium situation conforming to the changed conditions is basic. A case of this type has been described by Hutchinson (1967) in connection with the degradation of cliffs in Essex and Kent. This degradation proceeds in the form of progressive landslides.

Particular alterations of a coast may be triggered by an especially high flood, severe weather or similar events. This poses the problem of understanding the physical processes involved in such alterations in order to be able to design proper measures of protection. Fox and Davis (1973) have devised a simulation model for storm cycles and beach erosion in the case of Lake Michigan. The input data are meteorological ones (barometric pressure) in whose dependence the attendant changes in the near-shore circulation system and beach topography are determined and an empirical process-response model is set up. As noted, this model is entirely based on empiricism and not on hydrodynamic theory. It can be used for the prediction of expected changes due to various possible storm events on Lake Michigan, but does not give an insight into the physics of the phenomena involved.

As measures of protection, several general possibilities have been suggested by Jagschitz and Wakefield (1971); these include fencing and planting of beachgrass.

6.44. Tsunamis

When a large (Richter magnitude > 6.5) shallow earthquake occurs (depth less than 50 km) whose epicentre is in the ocean, it usually causes vast water-waves which can travel very far and cause diastrous effects on the coasts of the lands bordering that ocean. Such earthquake-caused water-waves are called "tsunamis". Many instances of the sudden arrival of such waves have been reported, particularly from the rim of the Pacific Ocean (see e.g. Adams, 1970).

The problem of elucidating the physics of tsunamis has several aspects. First, there is the problem of the generation of a tsunami by an earthquake, second the problem of the propagation of the disturbance over the vast distances observed, and third the problem of the action of the arriving wave upon the coast. Finally, the question of frequency of tsunamis in any one given tsunami-prone region is of interest.

Regarding the *generation* of the tsunami, it is generally believed that it is effected by the surfacing of an earthquake-fault at the ocean-bottom; in this, it is the dip-movement of the fault which activates the tsunami. Physically, the problem is therefore one of calculating the disturbance caused in an infinite, homogeneous and incompressible ocean of uniform depth H by an instantaneous displacement of a limited region of the sea-bottom represented by a step-function. The tools for solving this problem may be found in Stoker's (1957) book; an actual solution referring to the above problem was obtained by Ben-Menahem and Rosenman (1972). The result is a rather complicated expression for the far-field deep-water waves caused by the disturbance. A numerical procedure for obtaining solutions was devised by Hwang et al. (1972). Qualitatively, one obtains a series of more-or-less concentric wave-rings whose exact geometry is determined by the prevailing bottom topography. Under certain conditions, a high directionality may be generated for the wave amplitudes (Hwang et al., 1972).

Next comes the propagation of a tsunami-wave across the open sea. Van Dorn (1965) found that the tsunami-waves propagate in mid-ocean with a (group) velocity v of 234–237 m/sec, which corresponds approximately to the limiting formula for long water waves:

$$v = (gH)^{1/2} \tag{6.44-1}$$

for an ocean depth of H = 5,500 m. Since the total energy of the tsunami is constant, the amplitude of the waves diminishes with increasing distance from the source. The higher frequencies, however, propagate faster than the low ones, so that dispersion is observed.

The impingement of a tsunami-wave onto a coast is basically governed by the same physical processes as the shoaling of any wave. Because this process is described by non-linear equations, a bore of great height may develop in a narrow channel. The exact behaviour of the waves is determined by the bottom topography; tsunami-wave refraction diagrams have been obtained by Keulegan (1970) using a computer. Strange resonance phenomena of oscillations in bays can cause great damage. Numerical procedures for calculating coastal effects have been devised by Preisendorfer (1972). Typically, there are 3 to 5 large oscillations, after which the motion dies down. Maximum waves may reach coastal run-ups corresponding to wave heights of some tens of metres above normal sea-level. Such run-ups can have catastrophic effects.

Finally, attempts have been made at making statistical investigations of the frequency of tsunamis in a given area. One might expect a similar linear log-frequency–magnitude relationship for tsunamis as exists for earthquakes, provided that a suitable tsunami magnitude can be defined. For the latter, a quantity proportional to the logarithm (to base 2) of the maximum run-up on a (hypothetical) shore situated at 1,000 km distance from the epicentre of the earthquake may be taken (Iida et al., 1967). It is then found that the tsunami data do not show the expected linear relation between the magnitude and the logarithm of the frequency of occurrence (Adams, 1973). No similar risk charts as for seismicity can therefore be made.

6.5. Catastrophes on the ocean-bottom

6.51. General remarks

Catastrophes which are specific to the ocean-bottom are primarily represented by the occurrence of subaqueous slides. The apparent coefficient of friction is even much less in such slides than in catastrophic landslides because the solid material merges thoroughly with the water and forms a "turbidity current". Accordingly, subaqueous slides may travel a tremendous distance and wreak havoc in their path.

The classic case of a turbidity current occurred on November 18, 1929, when an earthquake triggered a turbidity flow; it was described by Heezen and Ewing (1952). Because of transatlantic cable breaks, the progress of the flow could be traced exactly. Other cases have been reported from the western New Britain Trench (Krause et al., 1970). Again, the currents were triggered by earthquakes. In other cases, however, turbidity currents were also found having been triggered by the natural collapse of unstable marine slopes (Bjerrum, 1971a) or by waves (Henkel, 1970). Particularly interesting are the high velocities that were observed in such currents (28 m/sec in the Grand Banks event; 14 m/sec in the New Britain Trench event). The distances travelled are also great: 650 km in the Grand Banks case, 260 km in the New Britain Trench case. In the observed currents, the bottom-slope decreased with distance from the source; the velocity decreased accordingly.

6.52. Physical mechanism of turbidity currents

Turbidity currents are the main agents affecting the morphology of the ocean-bottom. In this, they can have catastrophic effects, as evidenced by the succession of cable breaks after the Grand Banks earthquake. However, the same physical phenomenon is not only of importance in submarine accidents, but also in connection with powder-snow avalanches, volcanic nuées ardentes and possibly landslides.

The possible physical mechanisms of turbidity currents were recently reviewed by the author (Scheidegger, 1970a). Accordingly, there is no question that the suspension mechanism is caused by the turbulence prevailing in the water. There is still a problem, though, of explaining the lack of quick dissipation of this turbulence into the surrounding medium. Thus, turbidity currents show the characteristic phenomenon of autosuspension. In this, they are dynamically quite different from sediment-transporting rivers. In rivers, the bed load is moved by the fluid, whereas in turbidity currents the sediments are moving the fluid.

If a turbidity current is flowing for some time, it is usually considered as a river-flow analog. Thus, a Chézy-type relation is used to describe it:

$$\bar{u} = C\sqrt{hSg(\rho_s - \rho)/\rho_s} \qquad (6.52\text{-}1)$$

where $\bar{u}$ is the mean flow velocity, h the thickness of the current, S the bed slope, ρ_s the density of the suspension, ρ the density of the fluid and C some constant. This constant must be found empirically.

The head of a turbidity current behaves differently. Its thickness h_h is double that of the steady-state tail:

$$h_h = 2h \qquad (6.52\text{-}2)$$

For the velocity in the head $\bar{u}_h$ one has empirically:

$$\bar{u}_h = k_1 \sqrt{g \frac{\rho_s - \rho}{\rho_s} h_h} \qquad (6.52\text{-}3)$$

where k_1 is some constant varying from 0.4 to 0.7. This shows the interesting fact that $\bar{u}_h$ for the head of the current is largely independent of the slope S.

The problem of mixing at turbidity-current heads has been studied thoroughly by Allen (1971a) who was able to show that the head must be transversely "fingered out" (with clefts and lobes) to provide a steady progress for the current-head.

The above studies of turbidity currents are all semi-empirical. It is not easy to make meaningful experiments because the scaling conditions are difficult to satisfy. Recent studies along these lines have been made by Kuenen and Sengupta (1970), by Van der Knaap and Eijpe (1968) and by Komar (1970).

In all, it must be said that the mechanism of turbidity currents is not yet very well understood. Some phenomenological river-flow analog can be used for their qualitative description, but for the prediction of speeds, etc., we have to rely on various uncertain heuristic constants.

Chapter 7. **AIR CATASTROPHES**

7.1. Introduction

7.11. Severe weather

"Air" catastrophes are produced by "severe" weather: high winds, large pressure gradients and huge amounts of precipitation are the cause of damage and the death of human beings. According to Cressman (1969), there were 30 "killer storms" (i.e., storms which caused the death of 100 or more persons) between 1900 and 1969 in the United States alone.

Most of the severe-weather systems are represented, in some form, by vortices. These range from small "dust devils" (miniature whirls over a hot desert-surface; cf. Ryan, 1972) to tornadoes, hurricanes and cyclones. The order of magnitude of the kinetic energy contained in such systems is estimated in Table X. The

TABLE X

Energy of severe-weather phenomena
(After Battan, 1961)

Severe-weather phenomenon	Kinetic energy (Joules)
Dust devil	$4 \cdot 10^{7}$
Tornado	$4 \cdot 10^{10}$
Thunderstorm	$4 \cdot 10^{12}$
Hurricane	$4 \cdot 10^{16}$
Cyclone	$4 \cdot 10^{17}$

exact phenomenology of these features will be described when their physical aspects are presented.

Before examining the physics of severe-weather phenomena in

detail, it will, however, be necessary to present some basic physical principles regarding the atmosphere. Then, the severe-weather phenomena can be discussed.

7.12. Physics of the atmosphere

For the explanation of atmospheric catastrophes, only that layer of the air envelope of the earth which lies close to the surface is important. Nevertheless, a general understanding of fundamental meteorological processes is necessary in order to explain severe-weather phenomena.

Beginning with the statics of the atmosphere, we note that air can be treated in a first approximation as an ideal gas. Its equation of state is then:

$$\frac{p}{\rho} = RT \tag{7.12-1}$$

where p is pressure, ρ the density, T the absolute temperature (in degrees Kelvin) and R the gas constant divided by the molecular weight of the air. For dry air one has:

$$R \approx 2.87 \cdot 10^6 \text{ cm}^2 \text{ sec}^{-2} \text{ deg}^{-1} \tag{7.12-2}$$

The molecular weight of water vapour is only 18/29 of that of dry air; hence R becomes larger for moist air than for dry air. Under the same pressure and temperature conditions, the density of moist air is therefore *less* than that of dry air.

Equilibrium is attained in a vertical column of air if the following equilibrium condition is satisfied:

$$\frac{\mathrm{d}p}{\mathrm{d}z} = -\rho g \tag{7.12-3}$$

where z denotes the vertical coordinate (positive upwards) and g the gravity acceleration (which is a function of z). Inserting the value of ρ from eq.7.12-1 into eq.7.12-3 yields:

$$\frac{\mathrm{d}p}{\mathrm{d}z} = -g\,\frac{p}{RT} \tag{7.12-4}$$

Since this is a differential equation with separated variables, it can be integrated. One obtains:

$$p = p_0 \exp\left\{-\int_0^z \frac{g}{RT}\,\mathrm{d}z\right\} \tag{7.12-5}$$

or:

$$\rho = \rho_0 \exp\left\{-\int_0^z \frac{g}{RT}\,\mathrm{d}z\right\} \tag{7.12-6}$$

Since the distribution of g with height is independent of the atmosphere, the above equation yields a connection between the pressure (or density) distribution and the temperature distribution with height. This temperature distribution could, of course, be arbitrary; eq.7.12-5 or 7.12-6 would then simply yield the distribution of pressure or density corresponding to static equilibrium. However, a column of air can only be in gravitational equilibrium if, upon movement of a volume of air through a distance dz, the temperature change caused by the compression or expansion due to the pressure change with height corresponds to the actually prevailing temperature change. If one assumes that the temperature change dT in air is adiabatic, one has:

$$\mathrm{d}T = \frac{\mathrm{d}p}{p}\left(\frac{\gamma - 1}{\gamma}\right) \tag{7.12-7}$$

with $\gamma = 1.41$ for dry air. If a volume of air is moved through the distance dz, the pressure change is:

$$\mathrm{d}p = -g\rho\,\mathrm{d}z = -g\frac{p}{RT}\,\mathrm{d}z \tag{7.12-8}$$

Eliminating dp/p from eqs.7.12-7 and 7.12-8 yields:

$$\frac{\mathrm{d}T}{\mathrm{d}z} = -g\,\frac{\gamma - 1}{R\gamma} \tag{7.12-9}$$

This represents the temperature gradient which must prevail if the displacement of a volume of air upward or downward is not to produce a disequilibrium. It is called the "dry adiabatic lapse gradient" and is equal to roughly $-1°C$ per 100 m near the surface of the earth.

If the condensation of water vapour takes place during the ascent of a volume of air, heat is being released and added to it so that for condensing moist air the lapse gradient is smaller. The "moist adiabatic lapse gradient" is equal to about $-0.6°C$ per 100 m.

Stability, of course, is not always present in the atmosphere. If it is absent, convective "plumes", "thermals" and downdrafts result. These are vertical currents of air, originally arranged in a hexagonal cellular pattern (in plan) which may, however, become distorted into sheets if the air is also moving horizontally. Heating of the earth's surface (by insolation) is one of the most frequent causes of the origin of instabilities. It is these instabilities which cause many severe-weather phenomena.

7.13. Vortices

Pressure differences in a horizontal direction will result in an air flow. The complete explanation of the dynamics of the atmosphere is a complicated problem, but certain approximations lead to relatively simple results. If the vertical accelerations and the frictional forces are neglected, one obtains the *geostrophic* flow, which is a special type of flow.

This type of flow is subject to only two forces. The first is the Coriolis force F_c which acts normally to the wind velocity v, to the right in the Northern and to the left in the Southern Hemisphere. Its value is:

$$F_c = 2\omega v \sin \phi \qquad (7.13\text{-}1)$$

where ϕ is the latitude and ω the angular velocity of the earth.

The second force F_p affecting geostrophic flow is related to the horizontal pressure gradient. It acts normally to the isobaric lines. Denoting the pressure gradient normal to the isobars by $\partial p/\partial n$, one has:

$$F_p = \frac{1}{\rho} \frac{\partial p}{\partial n} \tag{7.13-2}$$

where ρ is again the density of the flowing air. In stationary flow, F_c must equal F_p; therefore, the geostrophic wind blows parallel to the isobaric lines with a velocity v:

$$v = \frac{1}{2\rho\omega \sin \varphi} \frac{\partial p}{\partial n} \tag{7.13-3}$$

Thus, if centres exist of high and of low pressure, around which there are closed isobars, a circular flow around these centres will result. The geostrophic flow theory only refers to frictionless, steady-state flow, and is therefore a gross oversimplification, but it shows, nevertheless, that circular motions of the wind are favoured. These circular motions represent *vortices* of various sizes. The friction which is neglected in the geostrophic wind theory causes the wind velocity to slow down and as a result there is a decrease in the Coriolis force. Hence, a small lateral velocity component *toward* the low-pressure region results. In a vortex around a low-pressure area, the wind tends to enter the centre, increasing its speed because of the conservation of angular momentum. For the conservation of mass, updrafts and downdrafts are then required. Vortices may not only be created by geostrophic rotation, but also by updrafts and downdrafts in an unstable atmosphere.

Vortices of various sizes are distinguished. The smallest are the dust devils, the next size are the tornadoes, then hurricanes and finally cyclones, as these features were mentioned in Sect.7.11.

7.14. The near-ground layer of the atmosphere

The near-ground layer of the atmosphere is of particular geomorphic significance. The effect of friction is that, as in any other liquid, a boundary layer develops near a contact surface. This boundary layer is turbulent; within it, immediately adjacent to the surface, is a laminar sub-layer.

The theory of boundary layers has been studied thoroughly in aerodynamics and in hydrodynamics. Accordingly, the distribution of the mean horizontal velocity $\bar{u}$ with distance y from a wall is (Kármán, 1930):

$$\frac{u}{u_*} = \frac{1}{k} \ln \frac{y}{y_0} \qquad (7.14\text{-}1)$$

where k is a universal constant equal to about 0.4, y_0 a constant of integration and u_* is the "shear velocity" given by:

$$u_* = \sqrt{\sigma_m / \rho} \qquad (7.14\text{-}2)$$

Here, σ_m is the maximum shearing stress occurring at the wall and ρ again the density of the air.

Kármán's law for the velocity distribution near a wall can also be written, if experimental values determined in a wind tunnel (Bagnold, 1941) are used for the constants:

$$\bar{u} = 5.75\, u_* \log_{10}(y/c) \qquad (7.14\text{-}3)$$

where c is a constant indicative of the surface roughness of the wall, equal to 1/30 of the diameter of the bumps causing this roughness.

The implications of the dynamics of the near-ground layer of the atmosphere form the subject of a special science which is called "micrometeorology". A standard treatise on it has been written by Sutton (1943).

7.15. Global circulation

It has been known for a long time that a mean global circulation exists. Mariners observed steady trade-winds; these were used to aid the sea passage between continents.

The normal distribution of atmospheric pressure has a high near the Pole, a low in the sub-polar regions, a high around 35° (doldrums) and again a low near the Equator (Fig.43). The consequence of this pressure distribution is that air rises over the Equator, flows poleward at great heights, descends in the doldrums, and then proceeds equatorward near the earth's surface, thereby creat-

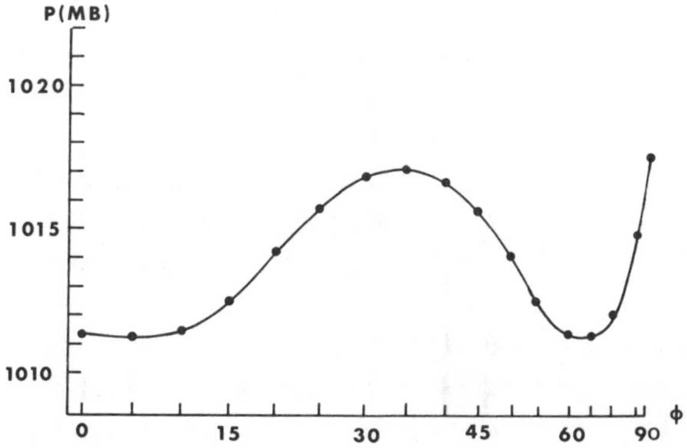

Fig. 43. Typical distribution of the zonally averaged sea-level pressure. (Modified after Pfeffer, 1964.)

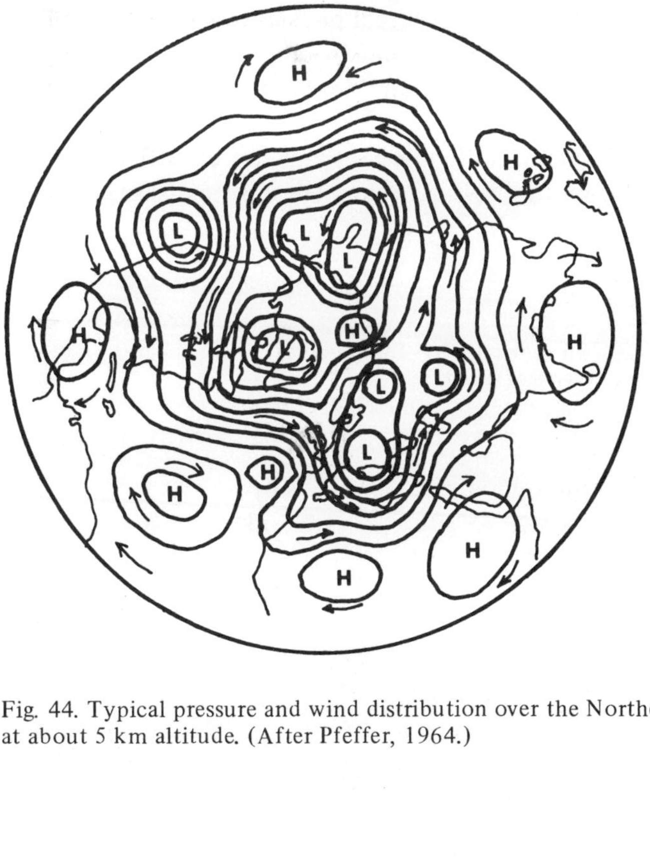

Fig. 44. Typical pressure and wind distribution over the Northern Hemisphere at about 5 km altitude. (After Pfeffer, 1964.)

ing the trade-winds. The Coriolis force would occasion the deflection of the winds from the north–south direction towards the east. These constitute the trade-winds. (Meridional cellular theory.)

The circulation outside the trade-wind zone is complex and not entirely understood. A typical pressure and wind distribution over the Northern Hemisphere at approximately 5 km altitude (after Pfeffer, 1964) is shown in Fig.44. A typical feature is the meandering jet stream encircling the globe.

It appears that the meridional cellular theory indicated above does not completely account for all the features of the circulation, and that the momentum transfer from easterly to westerly winds is the key to the understanding of it.

7.16. Clouds

Clouds arise if water vapour condenses and forms droplets or ice crystals in the atmosphere. The actual physics of clouds is not very important in relation to natural catastrophes, but the existence of clouds may be concomitant with the development of severe weather.

In the international classification, clouds are classified into several types, as shown in Table XI. The main distinguishing element

TABLE XI

Classification of clouds

Type	Air motion (vertical speed) (cm/sec)	Dimensions (km)		Precipitation
		vert.	horiz.	
Stratiform				
Lenticular	10^2	1	10	–
Stratus, Stratocumulus	$10–10^2$	1	$10^2–10^3$	drizzle light snow
Cirrostratus (multi-layered)	10	10	10^3	rain, snow
Cumuliform				
Cumulus	10^2	1	1	–
Cumulonimbus	10^3	10	10	showers

is whether clouds are stratiform (extending over large regions) or cumuliform (convection clouds).

Precipitation occurs when cloud particles reach such a size that they fall from the clouds. In this fashion, rain, hail or snow showers occur.

7.2. Thunderstorms

7.21. Phenomenology

Everyone has experienced thunderstorms: they are frequent occurrences of violent weather.

The small cumulus-clouds often seen in good weather may become unstable and develop into thunderheads (cumulonimbus-clouds). For this to happen, an unstable atmosphere must reach great heights, so that ascending air continues to rise due to a favourable lapse gradient. Upward air speeds in thunderclouds reach from 1 m/sec to 25 m/sec when the air reaches the 7,500-m level. Thunderstorm-clouds may reach heights of 12 km.

Byers and Braham (1949) introduced the concept of a "thunderstorm-cell", according to which a convection cycle arises in the thunderhead: when the updraft reaches such heights that water-vapour condenses, part of the cloud-air begins to descend owing to the weight of water and ice particles. Thus, one side of the thundercloud is ascending, the other part is descending. This stage of cell development is accompanied by heavy precipitation and electrical discharges. Finally, the downdrafts become dominant in the thundercloud and the latter begins to dissipate. The average life cycle of a thunderstorm-cell is about 30 minutes.

Whilst the precipitation from a thunderstorm can cause considerable damage by its quantity (the terminal velocity of raindrops is increased by a vertical effective stress, cf. Dawson and Warrender, 1973), particularly if a series of thunderstorms pass over one region in brief succession (25 cm of rain has been measured in a single day in parts of Illinois), it is the electrical effects of a storm with which we are mostly concerned here. Of these, lightning is of greatest importance. It is a spark-like discharge which can cause considerable damage. The most common form of lightning occurs in the form of an avalanche of charges proceeding from the cloud towards the earth in a series of steps of approximately 100 m each. When this charge-avalanche reaches the ground, the actual discharge takes place in the form of a return stroke. Other forms of lightning occur in the form of a single stroke from the ground up, particularly from high mountain peaks, and in the form of ball lightning.

The damage caused by lightning is due to the considerable electrical currents associated with it. Trees may be split, livestock and people may be killed, and fires may be started in buildings or in forests (Fuquay et al., 1972).

In order to understand the physics of lightning, it is first of all necessary to discuss the general principles of atmospheric electricity. Then, the specific questions of the generation of thunderstorm charges must be analyzed and, finally, the physics of the various types of lightning discharges can be investigated.

7.22. Atmospheric electricity

The atmospheric electricity is part of a world-wide circuit system. The basic pattern of this system was established by Wilson (1920). The pattern proposed by Wilson is shown in Fig.45. Accordingly, the ionosphere constitutes approximately an equipotential surface at a tension of approximately E_i = 300 kV (150 kV to 450 kV) relative to the ground. Systematic variations are caused by

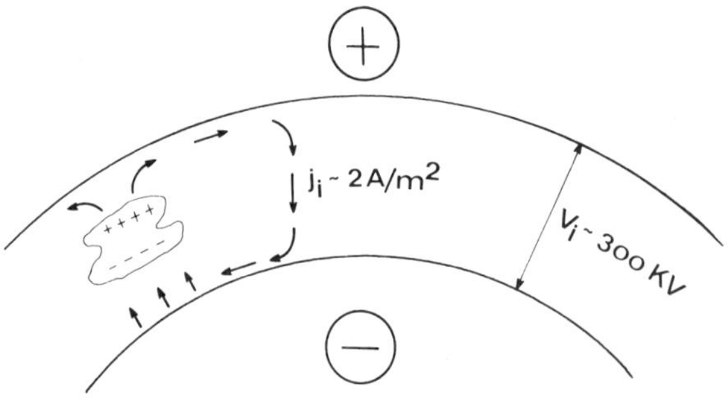

Fig. 45. Wilson's (1920) atmospheric electricity system.

solar-diurnal effects, while solar flares form outright disturbances (Bossolasco et al., 1972). There is, therefore, a voltage gradient in the atmosphere at all times and, because of the conductivity of the air caused by ions contained within it, a corresponding (fair-weather) electric current j_i. The latter is (Mühleisen, 1957) on land about $j_i = 2.3$ A/m^2 and at sea (in 1970) $j_i = 0.9$ A/m^2. There has been a steady decrease in the atmospheric current over approximately the last 50 years because of a diminution of the ion content of the air owing to increasing pollution.

The fair-weather current is compensated by electrical discharges in thunderstorms. In each discharge, the charge destroyed is about 20 to 30 C, the corresponding electric moment destroyed is about 100 C·m (Mason, 1972). It is one of the major problems in the science of atmospheric electricity to explain the mechanism of charge separation in the times required in thunderstorms.

7.23. Generation of thunderstorm-electricity

In spite of many research efforts, the mechanism of the generation of thunderstorm electricity is not yet entirely understood.

Measurements (Simpson and Scrase, 1937) have yielded the result that the charges in a thunderhead are distributed as shown in Fig.46. As noted earlier, the average duration of the life cycle of a

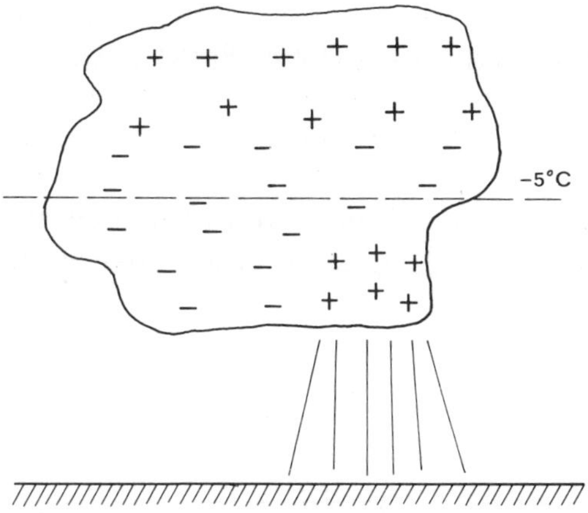

Fig. 46. Charge distribution in a thundercloud.

thunderstorm-cell is perhaps 30 min, the charge destroyed in each lightning flash is 20–30 C. This charge must be generated and separated in the region of the thunderhead which has a radius of about 2 km. The main negative charge is situated near the –5°C isotherm, positive charges exist some kilometres higher and are also near the bottom of the cloud.

Any sensible theory of the generation of thunderstorm-electricity must account *quantitatively* for the observed features listed above. Unfortunately, no such theory has yet been found. Of the theories proposed in the literature, we mention the following.

Wilson (1929) and Frenkel (1947) based their theory on the selective capture of negative ions by polarized raindrops. However, Mason (1972) noted that this mechanism is not capable of producing fields stronger than 500 V/cm, which is too small for initiating thunderstorm-activity.

Simpson (1909) proposed a charge-separation mechanism which occurs due to the break-up of water drops; this mechanism also appears to be much too weak to initiate thunderstorm-activity.

Pühringer (1961) assumed that electromagnetic induction in a cloud caused by the magnetic field of the earth would be sufficient to separate the charges.

Workman and Reynolds (1948, 1950) advocated that charge separation occurs when the water-drops are shed from the surfaces of large wet hailstones. In this fashion, an electrical tension between water and ice would arise.

Latham and Mason (1961) proposed that a potential difference could occur in the ice itself. Thus, the mechanism of charge generation would depend essentially on the development and electrification of hail-particles.

As can be seen from the diversity of the theories proposed, the problem of the generation of thunderstorm-electricity has by no means been solved yet.

7.24. Lightning

General remarks

Lightning represents a discharge between electrically charged atmospheric layers and/or the ground-layers. Thus, lightning may occur within a cloud or between clouds and the ground. In connection with "catastrophes", only the latter type is of interest.

Research into the physics of lightning has been conducted, particularly in connection with the protection of electrical power developments; much of what is known on the subject has been summarized in a monograph by Uman (1969). There are basically three types of lightning discharges that are encountered: those that proceed from the clouds to the ground – which is the most commonly encountered case; those that proceed from the ground upwards; and more unusual types such as ball- and bead-lightning. We shall discuss these various types in turn below.

Downward cloud-to-ground discharges

In flat country, the normal evolution of a thunderstorm-cell includes lightning which proceeds from the cloud to the ground.

As the charge generation and separation increases in the thundercloud, values for the electrical field are reached in which a discharge avalanche towards the ground begins to develop. This may begin from negative or positive charge-bearing sections of the cloud. The normal case is that a negative charge is lowered from the cloud to the ground. The breakdown voltage for ionization in air is known to be approximately $3 \cdot 10^6$ V/m. This value must be reached at the peak of the charge avalanche, although the measured field strength in thunderclouds is only of the order of 10^5 V/m.

The charge avalanche proceeds downwards in steps of approximately 50–100 m, until it reaches the ground ("stepped leader"). In this process, the leading stroke is, so to speak, ignorant of the ground; only when it arrives approximately 100 m or so above the ground is it possible for the stroke to aim at a tree or another likely target.

The stepped leader generally travels towards the ground with average velocities reaching from approximately 10^5 m/sec to more than 10^6 m/sec. Pause times between steps are approximately 50 μsec. The velocity during a step must exceed $5 \cdot 10^7$ m/sec. The stepped leader may branch during its journey. Typically, it carries a charge of 5 C which gives an average current of 100 A. The whole mechanism of stepped-leader formation is not very well understood; Uman (1969) quotes about half a dozen theories, but not one is entirely satisfactory.

Once the stepped leader touches the ground, the main discharge takes place in the form of a return stroke which constitutes the actual lightning. There may be several return strokes along the channel; in this case, these are again preceded by leaders from the cloud to the ground; however, these leaders travel faster and more smoothly than the original stepped leader. They are called dart leaders. Such dart leaders are glowing sections of the channel and travel with velocities of $2 \cdot 10^6$ m/sec.

As noted, it is the return strokes which constitute the "lightning". Our knowledge of the manner in which the discharge occurs is mostly due to the measurements taken atop Mount San Salvatore in Switzerland (Berger and Vogelsanger, 1965). According to these measurements discharges may reach 80 C and have peak-currents of from several to (rarely) 200 kA. Usually the current increases very rapidly in the first 10–20 μsec, then there is a less rapid decrease to 20% of the peak-value in the following 200–300 μsec. A typical example of the measurements is shown in Fig.47.

The type of lightning described above represents the "normal" type. Sometimes, the downward-branching stepped leader carries a positive rather than a negative charge. Positive currents need a longer time for the peak current value to develop and decrease.

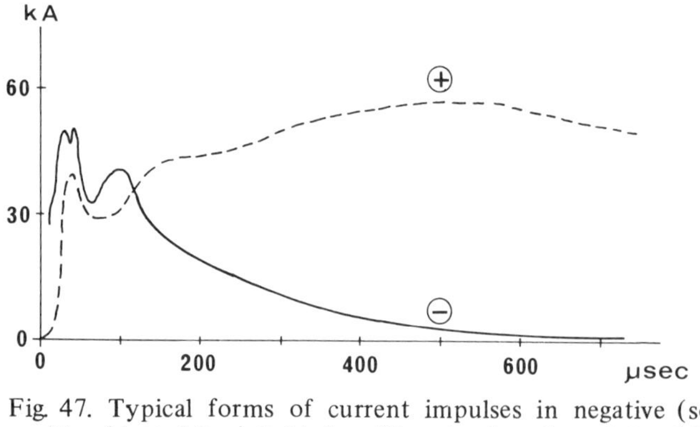

Fig. 47. Typical forms of current impulses in negative (solid line) and in positive (dotted line) lightning. (Drawing based on data of Berger and Vogelsanger, 1965.)

Equally, higher currents (a maximum of 218 kA was found by Berger and Vogelsanger, 1965) are observed. Similarly, the charges carried are also higher (up to 200 C). A typical example of the measurement results of a positive discharge is also shown in Fig.47. It was found that about 15% of the strokes initiated by downward-moving leaders carried a positive charge. Positive discharges do not appear to be of the multiple type.

Apart from the cases discussed above, a stepped leader may be formed from the ground upwards under very unusual circumstances (Orville and Berger, 1973).

For design purposes, it is of importance to have some statistical information about the frequency of various discharges. Fig.48 gives the frequency of peak-discharges observed in negative and positive downward strokes and Fig.49 the corresponding charge (modified after Berger and Vogelsanger, 1965), but it should be remembered that the positive strokes comprise only approximately 15% of the strokes observed. The data of Berger and Vogelsanger were obtained in southern Switzerland, but similar values were also observed in Florida (Uman et al., 1973a,b).

From these values an idea can be gained of the stress that may be placed on structures and living beings when struck by lightning.

Other phenomenological aspects of lightning refer to the fact

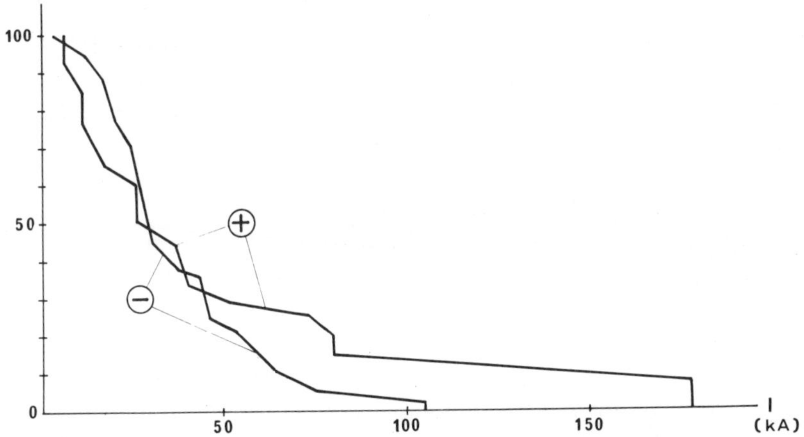

Fig. 48. Frequency of peak-currents in negative and positive downward strokes. (Modified after Berger and Vogelsanger, 1965.)

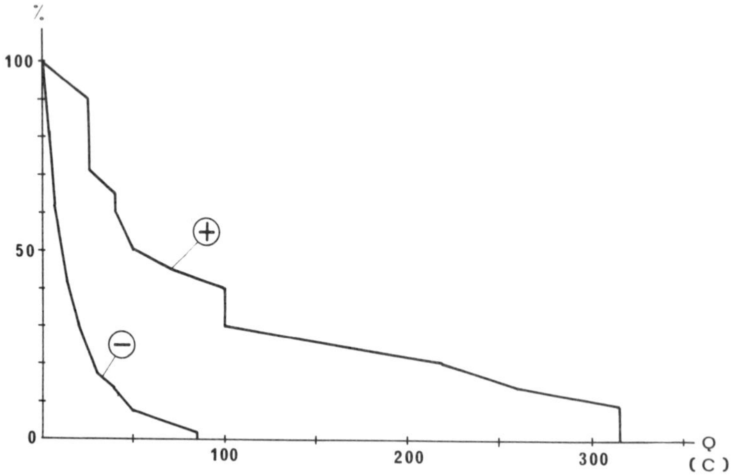

Fig. 49. Frequency of peak-charges in downward strokes. (Modified after Berger and Vogelsanger, 1965.)

that during the first few microseconds of a discharge, the channel is probably only about 1 mm thick. It then grows to about 1 cm (Berger, 1968). Sudden heating by the ionization causes thunder and it may split trees. Determinations of the temperature of lightning channels have been variously attempted. Orville (1968) found in observations of radiation values of 30,000–40,000°K. Calculations based upon a theoretical model of lightning also arrived at this value (Uman, 1969; Hill, 1971). This model also provides a reasonable explanation of the physical processes occurring during the return stroke which, in contrast to leader development, is understandable in terms of the behaviour of a plasma.

Upward ground-to-cloud lightning

Lightning between clouds and tall structures or mountain peaks occurs in a fashion different from that discussed above. Stepped leaders begin at the peak and work their way upward; they also branch upward. There is then no return stroke. Many observations of this type of lightning have been made on Monte San Salvatore in Switzerland (Berger and Vogelsanger, 1965) as well as on the Empire State Building in New York (McEchron, 1939). Again, it may be positive or negative charges that are being lowered from the sky to the ground. The general phenomenology and observational values for this type of lightning are similar to that observed in leader-induced lightning proceeding from the sky to the ground.

Bead- and ball-lightning

To conclude this section on lightning discharges we wish to mention some rare phenomena that have been observed.

Bead-lightning is the intermittent incandescence of a lightning channel in a pinch-and-swell effect: the channel breaks into luminous beads which seem to persist for a longer period than the usual lightning-discharge channel. Uman (1969) discussed some theories of this effect. As far as deletorious effects are concerned, only the main discharge is of importance.

A different (and as yet not very well understood) lightning-discharge phenomenon is ball-lightning: luminous balls, approximately 20–30 cm in diameter, which are associated in some way with thunderstorms, have been observed proceeding along irregular paths; they persist for several seconds, and then disappear silently or with an explosive effect. They are known to have killed people and to have caused appreciable damage. They are, however, quite rare. As noted above, the physics of the phenomenon is not very well understood. Singer (1971) has published a monograph on the subject in which he reviewed the many theories that have been proposed in the literature. These range from comparisons of ball-lightning with Leyden-jar phenomena to theories based on plasma models.

7.25. Lightning risk and protection

Introduction

Any evaluation of the risk of lightning damage at a location must begin with a climatological study: thunderstorms occur more frequently in some regions than in others, and thunderstorms produce the lightning which causes damage.

Frequency of lightning

According to the introductory remarks, one will have to begin with a statistical study of the frequency of thunderstorms with damaging lightning in an area. Usually, the number of days per year when such storms occur is noted. Further statistics can be compiled showing time of the year and time of the day when damaging lightning has occurred. Such a study has been made, for instance, by Chagnon (1964) for Illinois. In that part of the world, July is the peak-month for lightning and 13 h to 16 h local time are the peak hours at which it occurs. An interesting fact is that these are not the peak-hours for the occurrence of thunderstorms (20–23 h) which demonstrates that not all thunderstorms are equally severe. The incidence of damaging lightning can then be related to unit areas, giving a risk figure. Since the observation of "damaging" lightning depends on population density, allowance has to be made for fluctuations in this density in order to arrive at an unbiassed evaluation of the risk. Statistical studies such as that which has been made for Illinois by Chagnon (1964) are basic for any further statements on the damage that lightning may cause in an area.

Statistical data are also available for the discharges of thunderstorms in general. It was found in the middle latitudes (Montana:

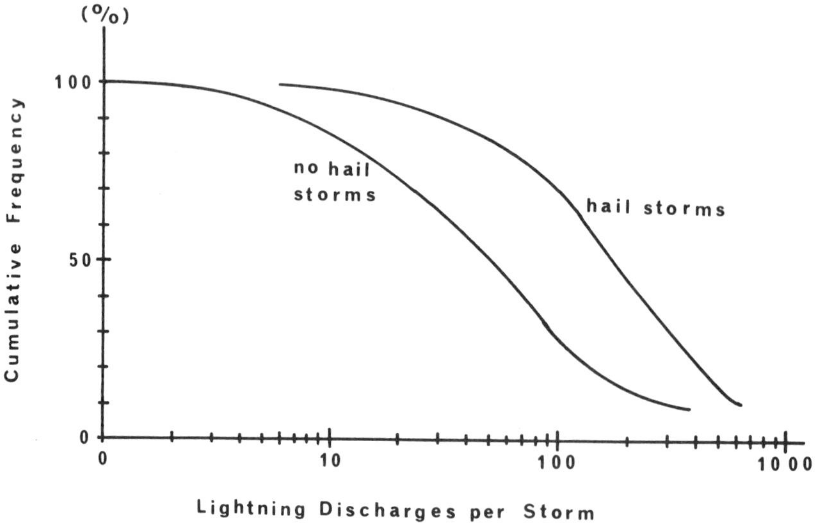

Fig. 50. Cumulative frequency of lightning-discharges per storm with and without hail in western Montana. (After Fuquay, 1967.)

Fuquay, 1967) that storms with hail produce many more lightning flashes than storms without hail. A single thunderstorm may produce up to 600 lightning flashes (Fig.50).

The physics of lightning damage

The mechanism by which lightning inflicts damage upon structures and living beings has to be sought in the high voltage and currents along the lightning channel. Typical current-pulses have been shown in Fig.47; temperatures may reach 40,000°K. If the channel enters a tree, the latter may split due to the vaporization of sap, or it may be ignited. Thus, lightning is a frequent cause of forest fires (Fuquay, 1967). Ignition is also possible in barns, etc. Trees with a rough bark (oaks) are more likely to be entered by the lightning bolt than trees with a smooth bark (beeches).

In people and animals, the high current through the body (in man the resistance from head to toe is about 500 Ω) is likely to set off ventricular fibrillations in the heart (Berger, 1971). It seems that, for short impulses (of less than 1 msec duration) a total charge of $Q > 0.1$ C flowing through the body represents the fibrillation threshold. The ground-to-sky lightning observed on mountain peaks, etc., is much more dangerous than sky-to-ground lightning. The number of fatalities due to lightning is not negligible. In Illinois, for example, an average of 6 people were killed annually in the period from 1926–1964, which is more than the number of people killed by tornadoes in the same state.

Protection

Before designing a lightning protection system, it must be decided whether the outlay is economically justified in the light of statistical data collected according to the principles described earlier. If it is decided that lightning protection is economically justified, an attempt will have to be made to provide for the harmless grounding of the current pulses which are expected when lightning strikes. In essence, this means that the structure to be protected is provided with a Faraday-type shield. The use of lightning-rods is a common means of protection; a lightning-rod is supposed to attract the stepped leader when it reaches the rod's vicinity; thereby protecting a cone of 90° angle below the vertex.

The action of a lightning-rod, however, is not as simple as it may appear from the above remarks. The problem has been investigated by Schwab (1965) and by Golde (1967). Accordingly, the electrical field in the vicinity of the head of a stepped leader can be calculated; it depends on the charge of the leader. This charge is distributed over the entire length of the leader; therefore the electrical field strength at its head depends on the height H of its origin ("height of the thundercloud") as well as on the charge Q. A stroke will occur if the field strength between the head of the leader and a grounded point exceeds the breakdown field strength of 3 kV/cm in the air. Thus, one can calculate the "striking distance" h for lightning strokes from the head of the leader to a grounded point as a function of H and Q. For practical applications, the dependence of h on peak-current i is of more interest than the dependence on the charge Q. Assuming the half-duration time of an average lightning discharge to be 50 μsec, one obtains for the peak-current:

$$\hat{i} = 2 \cdot 10^4\, Q \tag{7.25-1}$$

Thus, one can calculate the striking distance h for lightning

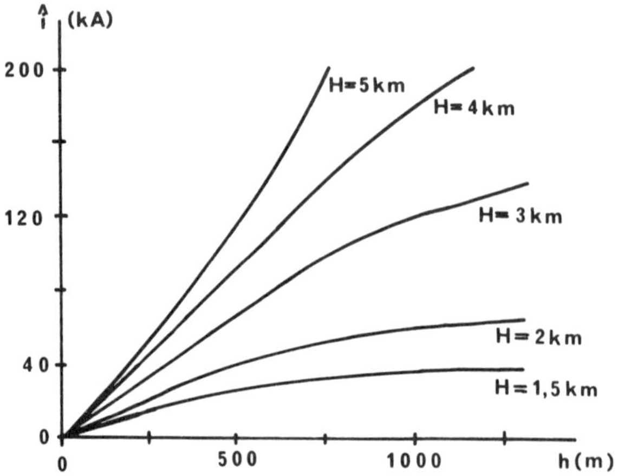

Fig. 51. Striking distance h of a lightning bolt of discharge current $\hat{i}$ in dependence of various thundercloud heights H. (After Schwab, 1965.)

strokes of various discharge values. This distance is the distance from which a grounded conductor can "attract" a lightning stroke. The results of the pertinent calculations are shown in Fig.51. This figure provides all the means for the design of a lightning protection system, inasmuch as the region from which a lightning bolt of a given discharge can be attracted, can be determined. It appears that grounded roof wires may often provide better protection than a series of lightning-rods.

A person caught in the open when a thunderstorm occurs should seek grounded shelter; a forest may present such a shelter. Locations from which ground-to-sky discharges may begin, such as exposed mountain peaks or ledges, are extremely dangerous; to shelter near a single tree may also be dangerous because the tree acts as a lightning-rod and the danger exists that spark-shortening between the person and the tree will occur; the resistance of man (mentioned above as about 500 Ω) is likely to be less than that of the tree. If no shelter is available anywhere, the lightning-rod effect that a man has will be minimized by squatting on one's heels in some low-lying area.

7.3. Tornadoes

7.31. Phenomenological description

Tornadoes are violent vortices that have the appearance of variously shaped funnels which dip down from the base of existing clouds. They approach the ground in an irregular fashion and cause severe damage wherever they touch down. The damage is due to the high wind speeds prevalent in tornadoes as well as to the low pressure in their centres: pressure differences of up to 80 mbar between the core and the outer part of a tornado have been reported.

The damage caused by a tornado can be directly correlated with the maximum wind speed occurring therein. Upon this basis Fujita (1971) developed a damage scale of five classes. This scale (F = Fujita scale number) can be expressed by the formula:

$$v = 6.30(F+2)^{3/2} \text{ m/sec} \qquad (7.31\text{-}1)$$

where v is the wind velocity in m/sec. Accordingly, one has the following correspondences:

$F = 0$	$v =$ 18 m/sec
$F = 1$	$v =$ 33 m/sec
$F = 2$	$v =$ 50 m/sec
$F = 3$	$v =$ 70 m/sec
$F = 4$	$v =$ 93 m/sec
$F = 5$	$v =$ 117 m/sec

Fujita (1971) published a set of pictures illustrating the damage relating to each F number.

Tornadoes are actually quite small events. They are rarely larger than 1.5 km in diameter; many have a breadth of less than 100 m. Some may never reach the ground. Similarly, the ground-path length of tornadoes may be only a few hundred metres, although that of the more diastrous ones is known to have been hundreds of kilometres. A tornado travels along its path at an average speed of approximately 18 m/sec, but may reach 27–31 m/sec. The path length L and the diameter W are therefore two important parameters of a tornado. Since these quantities vary widely, it appears reasonable to introduce again a scale which also reaches from 1 to 5. This scale is chosen as a logarithmic one. For the length scale (P_1) one sets:

$$L = 1.61 \cdot 10^{(P_1 - 1)/2} \qquad (7.31\text{-}2)$$

and for the width scale (P_w):

$$W = 1.61 \cdot 10^{(P_w - 5)/2} \qquad (7.31\text{-}3)$$

This yields the following correspondences.
For the lengths of tornado paths:

$P_1 = 0$	$L =$ 0.5 km
$P_1 = 1$	$L =$ 1.6 km
$P_1 = 2$	$L =$ 5.1 km
$P_1 = 3$	$L =$ 16.1 km
$P_1 = 4$	$L =$ 50.9 km
$P_1 = 5$	$L =$ 161 km

For the widths of tornado paths:

$P_w = 0$	$W = 0.005$ km
$P_w = 1$	$W = 0.016$ km
$P_w = 2$	$W = 0.051$ km
$P_w = 3$	$W = 0.161$ km
$P_w = 4$	$W = 0.509$ km
$P_w = 5$	$W = 1.610$ km

The three numbers F, P_1, P_w represent a phenomenological characterization of a tornado.

Individual tornado events have been described on many occasions. A notable book has been authored by Flora (1953). Fujita (1959) has analyzed the Fargo tornadoes in North Dakota of June 20, 1957. A very thorough phenomenological study of approximately 30 tornadoes that occurred on Palm Sunday (April 11) 1965 in the American Midwest (causing 258 fatalities) has been published by Fujita et al. (1970). From such reports, the following phenomenological features of tornadoes have been abstracted:

(1) Tornadoes dip from the sky toward the ground; when they "touch down" they cause damage.

(2) The tangential tornado circulation is represented by a core (the eye) surrounded by a vortex. Ground-wind speeds may reach 90 m/sec.

(3) There is a consistent large pressure difference between the core and the outside of a tornado. The value in the core is 20 to 80 mbar lower than in the outside.

(4) There appear to be individual "suction spots" which rotate around the eye of the tornado with the speed of the rotational circulation velocity. The suction has lifted railway cars weighing 13 tons from their tracks. This would correspond to an upward wind speed of 100 m/sec (Pühringer, 1963).

(5) There is evidence that rapid downdrafts also exist within the reach of a tornado. Building wood has been slammed 45 cm into the ground after the passage of a tornado (Rossmann, 1959).

(6) Intense lightning and St. Elmo's fire is often (but apparently not always; cf. Rossmann, 1959) observed in conjunction with tornadoes.

(7) The speed with which the centre of a tornado travels along the ground is of the order of 27 m/sec.

7.32. Physics of tornadoes

Tornadoes are phenomena associated with characteristic synoptic conditions. As noted, Fujita et al. (1970) made a study of the conditions that produced the tornado outbreak on April 11, 1965 in the United States. It was found that, behind a cold front advancing eastward towards a mass of warm, moist air, there was a large tongue of cloud-free dry air. This appears to be the typical synoptic pattern favourable for the origination of tornadoes: dry cold air rapidly runs over a mass of warm moist air; it thereby creates a superadiabatic lapse rate within the warm air just ahead of the cold front on the ground, and causes instability in the atmosphere. Showalter (1953) proposed a stability index *SI* which is computed by lifting a volume of air at the 850-mbar surface first dry-adiabatically to saturation and then moist-adiabatically to the 500-mbar surface and taking the difference of the temperature of the thus lifted volume and the actual temperature prevailing at the 500-mbar surface. Conditions are stable when this stability index SI is $> +6°C$. When $SI < -4°C$, unusually severe storms may be expected.

The physics of the actual tornado vortex is not yet completely understood. Any physical theory has to account for the observational facts listed in Sect.7.31. Reviews of the various theories have been provided by Markgraf (1961) and by Brooks (1951). Accordingly, it was originally thought that a tornado is nothing but a thermal vortex created by rapidly rising air. However, the fact that tornadoes dip *down* from the clouds and may contain downdrafts, speaks against this model. The power of a tornado has been estimated to be approximately 3,500 MW (Pühringer, 1963), by calculating the kinetic energy per unit time that has to be supplied to move the tornado column of air upward at the observed speeds. It is very difficult to supply this power by thermomechanical means. Vonnegut (1960) has shown that the temperature differences to maintain tornadic wind speeds of 200 m/sec would have to be 70°C; for winds of 100 m/sec, 20°C. Such large temperature differences would have to be maintained, which is difficult to envisage from insolation. They can occur in volcanic eruptions and in large fires; these have been known to create whirlwinds.

Small vortices, such as dust devils, can also be explained by the above thermal model.

Modifications of the above thermodynamic model have been attempted by Showalter (1943) and Koschmieder (1940), on the basis of temperature inversion and moisture effects.

The inverse, so to speak, of the above theories is that a downward instability exists which dips from the cloud towards the ground. Then, the circulation would be *downward* in the eye of a tornado, upward from the edge of the eye on outward (Letzmann, 1921, 1923). The formation of hail could cause such an instability (Rossmann, 1959). Of course, similar temperature differences (20°–70°C) as in the reverse case must be required to supply the power requirements of a tornado. However, this is perhaps easier to envisage by a supply of falling hail than by insolation.

The difficulty of finding the power supply for the maintenance of a tornado has led to the consideration of the possibility that tornadoes are maintained by electrical forces. This idea is very old (Lucretius, 60 B.C.). A review of its history has been given by Vonnegut (1960). If the electrical energy in a thunder cloud causes 10–20 lightning strokes per second, then this would produce 10^8 kW of power, enough to sustain a tornado. This is, of course, a great amount of lightning. In some tornadoes, no electrical effects appear to have been noted at all; but perhaps they were missed by the eye-witnesses. No electrical effects seem to occur in waterspouts, which appear to be tornadoes "at sea". Thus, it is difficult to envisage that tornadoes are electrical effects.

A further problem is to find the mechanism by which the electrical energy would be transformed into the mechanical energy of the blowing wind. Vonnegut considers two possibilities. The first (Lucretius, 60 B.C.) is that the lightning would heat the air so that an intense thermal vortex would ensue; the second is that ionized air could be accelerated in a strong electrical field. The first method, accordingly, is again using thermal arguments. The notoriously low efficiency of thermal "engines" (in the present case perhaps 40%) makes it very difficult to obtain sufficient power to maintain the wind. In the second type of mechanism, a direct transformation of electrical into mechanical energy occurs. Pühringer (1963) in particular has followed up this possibility. However, a constant discharge of 10–20 lightning strokes (see above) would be required to maintain the tornado winds.

Therefore, it seems that none of the theorics of tornado processes advanced so far is very satisfactory. The numerical estimates lead to values which are perhaps more realistic in those theories that assume a downdraft instability originating in the clouds than in the other theories.

7.33. Statistics of tornadoes

Tornadoes are individual events of various scales and magnitudes. It is difficult, therefore, to find an easy means for the representation of the statistics of their occurrence.

Tornadoes occur in many parts of the world; they have been reported from Europe (Wegener, 1917), Australia (Clarke, 1962) and Japan (Fujita, 1971). However, in these regions they are only occasional and infrequent events. The region where they are most predominant and where they are an ever-present menace is the

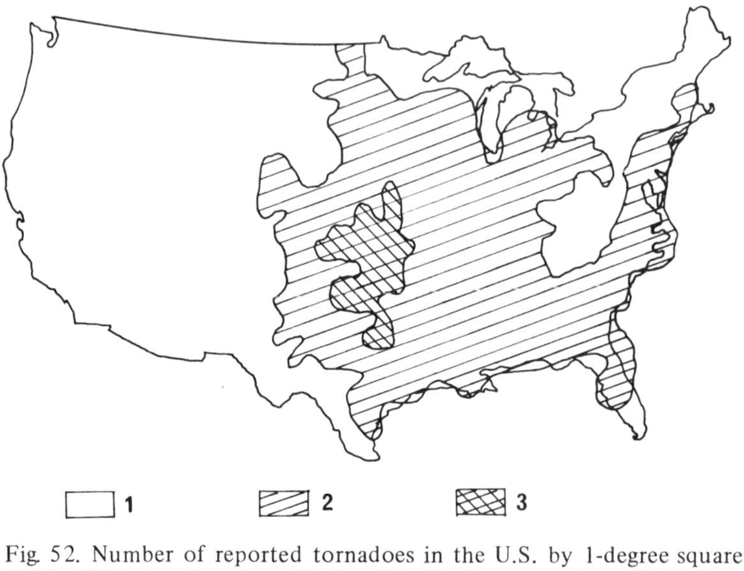

Fig. 52. Number of reported tornadoes in the U.S. by 1-degree squares from 1916–1961. *1* = less than 10; *2* = 10–50; *3* = more than 50. (After Wilson and Chagnon, 1971.)

Midwest of the United States of America (Flora, 1953): in this area, an average of more than 200 people are killed by them every year (Battan, 1961). Most of the discussion here will therefore be with reference to the United States.

Any statistics of tornadoes must begin with observations. The latter are much affected by population density, alertness and availability of observers, etc. Nevertheless, by collecting reports from various areas, an impression of the relative frequency of tornado incidence can be derived. Fig.52 gives the results for the tornadoes reported in the United States from 1916–1961 by 1-degree squares, as drawn from data published by Wilson and Chagnon (1971).

Tornadoes do not occur at random in time. They are (in the U.S.A.) more frequent in the spring than at any other time of the year. They are concentrated on tornado-*days*. Usually, a tornado-day has 2 tornadoes, but one tornado-day in Illinois had 19. Evidently, tornado events tend to occur in clusters. Therefore, a simple Poisson distribution does not adequately describe their annual frequencies in a region. Thom (1963) has shown that a negative binomial distribution is more adequate than a Poisson distribution to describe annual tornado frequencies. A negative binomial distribution $f(t)$ has the form:

$$f(t) = \frac{\Gamma(t+k)\, p^t}{\Gamma(t+1)\,\Gamma(k)\,(1+p)^{n+t}} \tag{7.33-1}$$

where p and k are parameters. Based on this distribution, Thom (1963) arrived at a method for establishing a predictive probability with which a point may be struck by a tornado within a year.

Unfortunately, the number of tornadoes occurring in an area is not necessarily an indicator of the probability of the tornado damage and risk that is to be expected there. Most of the diastrous effects from tornadoes have been caused by a few rare catastrophic storms. A tornado affecting parts of Missouri, Illinois and Indiana on March 18, 1925, ranks as one of the most devastating on record. It had a continuous path length of 352 km on the ground surface, a mean speed of 27.7 m/sec, and was from 0.8 to 1.6 km in width so it destroyed a total area of 425 km^2. It killed 695 people. A similar, although not quite as devastating, storm occurred in central Illinois on May 26, 1917 (length 472 km, average speed 18 m/sec, width 0.4–0.8 km, 103 people killed).

Since catastrophic tornado events are rare, it is difficult to arrive at a statistical method for their description or at a prediction of the probability of their occurrence at a particular location. An inspection of Fig.52 shows that Illinois does not belong to that area of America which is affected by the greatest number of tornadoes; it lies on the eastern fringe of the main tornado belt. Nevertheless, as noted, it experienced some of the most devastating tornadoes that have occurred anywhere. Therefore, up to the present time, nothing more can be done than to assume that tornadoes can occur again in the general regions where they have previously occurred, and to take precautions accordingly.

Rather than general predictions, specific ones ("tornado warnings") can be made when the atmospheric conditions are monitored, and particularly, if a tornado has already been spotted. In this fashion, it is often possible to warn the population in the likely tornado path to proceed to adequate shelters.

7.34. Physics of tornado damage

The damage a tornado causes to an enclosed structure is essentially due to two effects: the direct wind-pressure caused by the high wind speeds and the explosive effect caused by the rapid pressure-drop in the eye of the tornado (Melaragno, 1968; Fujita, 1971).

The forces due to wind velocity are of two types. Where the house-wall faces the wind, the deflection of the wind caused by the wall is experienced by the latter as a pressure. The pressure is proportional to the momentum change per unit time; the air momentum is ρv, the advection per unit time is proportional to v, hence the pressure p is:

$$p = kv^2 \qquad (7.34\text{-}1)$$

where k is a constant. If p is in kp/m^2, v in m/sec, k is found between 0.06 and 0.35 with 0.1 a good average value. For a wind velocity of 67 m/sec, one obtains a pressure of 449 kp/m^2 (Tannehill, 1952, p.123).

At the roof and in the lee of a structure a lifting force is experienced (airplane-wing effect).

There is a low-pressure region in a tornado's eye. When the eye passes quickly over a house, the pressure within the house cannot be immediately equalized, thus the latter will "explode". For a sudden pressure drop of 30 mbar, the outward acting explosive pressure would be equal to that amount. Naturally, the pressure that is actually experienced depends on the temporal course of the pressure-drop outside the house during the passage of the tornado and the amount of partial equalization that can take place.

There is no question that houses could be constructed in a manner which would make them tornado-resistant. The proof lies in the fact that heavily reinforced concrete buildings rarely suffer much damage. However, the buildings in many of the world's tornado areas are of light construction and building codes for tornado-proofing houses are absent; thus people have to rely on tornado warnings to seek shelter at the appropriate time. The best shelters are properly built "cyclone cellars" (as they are called in the Midwest of the U.S.); failing this, a sturdy part of the basement of a house will afford some protection. In the open, crouching in any depression or hole is better than remaining upright.

7.4. Hurricanes

7.41. Description

Hurricanes (sometimes called typhoons) are large atmospheric vortices with wind speeds of above 100 km/h; they develop in the doldrums of the tropics and move in an often erratic way towards higher latitudes. In the Northern Hemisphere they rotate anticlockwise, in the Southern Hemisphere their rotation is clockwise. Maximum windspeeds that have been recorded were approximately 64 m/sec (for one minute at San Juan, Puerto Rico, on September 13, 1928; see Tannehill, 1952). Typically, the rapid tangentially blowing winds of a hurricane enclose an area of calm called the *eye*. Eyes of hurricanes average 24 km in diameter but may reach 64 km. The edge of the eye is formed by a cloud wall. In the eye of a hurricane, the pressure is lower than outside it; a drop of 34.5 mbar has been recorded in Formosa on September 16–17, 1912.

The effects of hurricanes are caused by their high winds and by the strong precipitation associated with them. The banks of cloud and rain in a hurricane form spiral arms which begin at the edge of the eye. They form and dissipate rapidly. Huge storm-waves (up to 13 m above normal tide) may be created at sea by the high wind speeds; these waves can cause great damage when hitting coastal areas (Tannehill, 1952). Thus, the inundation of Galveston Island (Texas) caused by a hurricane in 1900 killed 6,000 people (Cressman, 1969). The heavy rains accompanying hurricanes can cause flash floods on land areas: 50 cm of rain dumped at a particular spot in half a day or so is not an uncommon occurrence (Battan, 1961). An extreme rainfall of 253 cm by hurricane Agnes on Mount Mitchell (N.C.) in less than 48 h from June 19 to 21, 1972 was reported by the U.S. National Weather Service (1972).

7.42. Physics of hurricanes

As is evident from their description in the last section, hurricanes represent giant vortices which may develop in the tropics under certain synoptic conditions and which then proceed along varied paths into higher (northern or southern) latitudes. If they strike land or reach middle latitudes, they slowly dissipate.

Thus, hurricanes are formed in tropical regions and are probably caused by some instability in the atmosphere. A supply of warm, moist air seems to be essential for their formation. Upward air motion must occur which will lead to condensation in the upper levels of the atmosphere and act as a trigger for storm centre formation. Some authors assume a convectional (gravitational) instability (Haque, 1952), a barotropic (inertial) instability (Alaka, 1958), or a baroclinic instability (Kleinschmidt, 1951; Yanai, 1961). Riehl (1954) supposed that the simultaneous arrival of a low-level and a high-level disturbance in the wind field at a warm point in the ocean would create the required instability. In most of these theories a pre-existing low is assumed to be essential for spawning a storm. Scorer (1966) has based his theory on the random occurrence of thermal disturbances, an idea which was further developed by Shabbar (1969) who allowed for a variable friction term and generally set the model on a quantitative basis.

In spite of the above attempts, the mechanics of hurricane formation is not very well understood. In general terms, one can say that there must be a continuous source of energy to start and maintain the hurricane. From the observation that hurricanes generally dissipate over land and in higher latitudes, the energy source must be sought in the moist air which rises from a warm ocean. The heat released by the condensation of the water supplies the energy for the hurricanes.

7.43. Physics of hurricane damage

The damage by hurricanes is due to three effects. First, the large wind velocities, second, the floods caused by high precipitation, and third, the possibility of storm waves.

The wind-pressure effects are similar to those experienced in tornadoes, only usually somewhat less, because the wind velocities are generally not as high as those encountered in tornadoes. The formulas have already been presented in Sect.7.34. Similarly, the missile effect of objects carried by hurricane-winds is akin to that in tornadoes; the pertinent expected wind velocities have to be used in the corresponding calculations. The explosive effect on houses caused in tornadoes by the low pressure at the centre of the vortex is not as noticeable in hurricanes, because the pressure variations occur in hurricanes more slowly than in tornadoes, so that equalization inside and outside a building can take place.

A severe danger not present in tornadoes is caused by the flash flooding due to the high precipitation prevalent in hurricanes. The effects of 50 cm or more of rain on a watershed can be predicted by the usual hydrograph theory. Corresponding estimates can be made.

The greatest damage and loss of life encountered when hurricanes occur, however, is due to storm waves impinging on coastal areas. The effect is similar to that of a tsunami. The progress of such a wave has to be calculated according to standard flood wave theory.

7.44. Hurricane statistics, risk and protection

As noted above, hurricanes are spawned in the doldrums and then move to higher latitudes. Although the path of an individual hurricane may be quite irregular, the average paths of hurricanes show a pattern that is consistent over the world. Fig.53 shows the regions of the world which experience hurricanes and the average

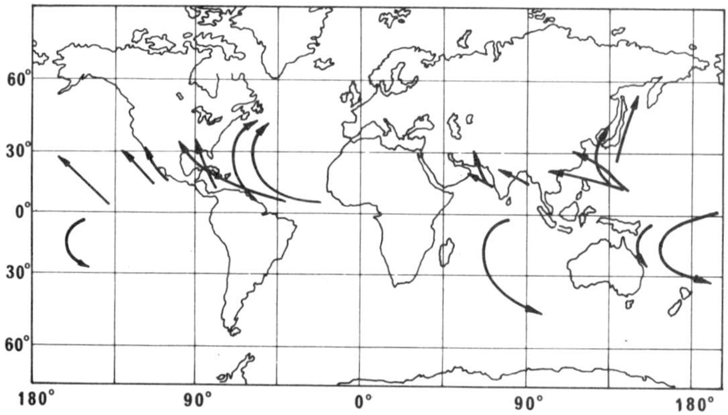

Fig. 53. Principal world regions affected by hurricanes and average directions of their movement. (Modified after Tannehill, 1954.)

paths of these hurricanes. Thom (1960) has made an annual-frequency analysis of hurricanes affecting the eastern United States and found that most of these frequencies follow Poisson distributions for which the variance is proportional to the mean in sampling:

$$f(x) = \mu^x e^{-\mu}/x! \qquad (7.44\text{-}1)$$

Here, $f(x)$ is the probability distribution, x the annual frequency and μ the mean annual frequency. For *all* hurricanes reaching the U.S. Atlantic coast, Thom (1960) found $\mu = 2.04$ and a mean square deviation of $\sigma^2 = 1.84$. For a Poisson distribution, one should have $\mu = \sigma^2$; Thom found by a chi-square test that the observed discrepancy is not significant and that it can be ascribed to the limitation of the available data series.

For most hurricane data series analyzed by him, Thom found that a Poisson distribution is adequate. For hurricanes observed at sea, he noted an upward trend in the present century which he ascribed to an improvement of observation techniques (more hurricanes were spotted in more recent times) so that, after adjustment for trend, a Poisson distribution again describes the data. It is reasonable to assume that similar annual-frequency analyses could also be made for other regions of the world which are affected by hurricanes.

The establishment of the probability distribution of hurricane incidence upon a region represents a prediction of the risk from hurricane damage that must be expected in that region over the years. Regarding the imminent danger from a specific hurricane, it is indicated to track its progress by radar airplanes and satellites and to try to estimate its course by means of short-range forcasting so as to give a warning to the threatened regions. In view of the possibility of great storm waves, floods and high winds, the only way to protect lives is often by evacuation of the threatened coastal areas. The damage to structures cannot be lessened in this fashion, but many lives can be saved.

7.5. Other types of severe weather

One could mention some severe weather types which do not fall into the categories listed above.

Winter blizzards in middle latitudes are nothing but the effects of cyclones (low-pressure centres) during cold weather. Large accumulations of snow combined with fairly high winds in their wake can paralyze all traffic and maroon travellers. The danger of blizzards is therefore rather an indirect one; if warnings are obtained in time, staying indoors and waiting until the storm has passed will provide all the protection needed.

Sometimes, when the snow melts rapidly late in the spring, floods can ensue. The probability of the occurrence of such spring floods can be learned from the statistics of floods.

In deserts, high winds can produce dust- and sand-storms. The physics of sand movement has been studied by Bagnold (1941).

Accordingly, there is a sharp upper bound for the height of a layer of blown sand which depends on the velocity of the wind. Highways can be quickly covered during a sand storm. The normal electrical field of the atmosphere can be affected (Kamra, 1972). The wanderings of sand dunes, which may be 20 m high, can also have catastrophic effects. By their inexorable progress they can engulf entire villages. Their wandering velocities of 3–5 cm per day are, however, slow enough for people to abandon their homesteads in time to save their lives. A dune moves by sand being blown from its windward "back" to the "slip surface" in the lee. Stabilization can be attempted by establishing vegetation on a dune.

REFERENCES

Achenbach, J.D. and Abo-Zena, A.M., 1973. *J. Geophys. Res.*, 78:866.

Adams, R.D., 1968. *Seismic Effects at Mangla Dam.* UNESCO Rept. Series No.975/BMS, Paris.

Adams, R.D., 1972. *Bull. Seismol. Soc. Am.*, 62:1787.

Adams, W.M. (Editor), 1970. *Tsunamis in the Pacific Ocean. Proc. Int. Symp. Tsunamis Tsunami Res.*, East-West Science Center, Honolulu.

Adams, W.M., 1973. *Proc. World Conf. Earthquake Eng., 5th, Rome*, Paper No.352.

Aggarwal, P.N., 1972. *Bull. Seismol. Soc. Am.*, 62:661.

Aggarwal, P.N., Sykes, L.R., Armbruster, J. and Sbar, M.L., 1973. *Nature*, 241:101.

Akademiya Nauk SSSR, 1961. *Zemletryasenii SSSR*. Akad. Nauk SSSR, Moscow.

Akademiya Nauk SSSR, 1970. *Fizicheskiye osnovaniya poishov metodov prognoza zemletryaseniy*, Akad. Nauk SSSR, Moscow.

Aki, K., 1968. *J. Geophys. Res.*, 73:5359.

Aki, K., 1972a. *Tectonophysics*, 13:423.

Aki, K., 1972b. *Tectonophysics*, 14:227.

Alaka, M.A., 1958. *Geophysica*, 6:133.

Alexander, G.N., 1972. *J. Hydrol.*, 16:225.

Alexander, S.S., 1968. *Earth Miner. Sci.*, 38(2):16.

Alfrey, I., 1944. *Q. Appl. Math.*, 2:113.

Allen, C.R., St.Amand, P., Richter, C.F. and Nordquist, J.M., 1965. *Bull. Seismol. Soc. Am.*, 55(4):753.

Allen, J.R.L., 1968. *Current Ripples*, North-Holland, Amsterdam.

Allen, J.R.L., 1971a. *J. Sediment. Petrol.*, 41:97.

Allen, J.R.L., 1971b. *J. Fluid Mech.*, 49:49.

Alsop, L.E. and Oliver, J.E., 1969. *Trans. Am. Geophys. Union*, 50(5):376.

Altmann, H., 1958. *Jahrb. Geogr. Ges. Bern*, 44:1.

Ambraseys, N.N., 1969. *Bull. (Izv.) Acad. Sci. USSR, Earth Phys.*, 1969(7):91.

Ambraseys, N.N., 1973. *Proc. World Conf. Earthquake Eng., 5th, Rome.* Invited Paper on "Dynamics of Response of Foundation Materials in Epicentral Regions of Strong Earthquakes."

Ambraseys, N.N. and Sarma, S., 1969. *Bull. Seismol. Soc. Am.*, 59:651.

Amein, M., 1966. *Water Resour. Res.*, 2(1):123.

Amein, M., 1968. *Water Resour. Res.*, 4(4):719.

Ampferer, O., 1939. *Sitz.-Ber. Akad. Wiss. Wien, Math.-Nat. Kl., Abt. 1*, 148:1.

Ampferer, O., 1940. *Sitz.-Ber. Akad. Wiss. Wien, Math.-Nat. Kl., Abt. 1*, 149:51.
Anderson, E.M., 1942. *The Dynamics of Faulting and Dyke Formation with Applications to Great Britain*. Oliver and Boyd, Edinburgh.
Anonymous, 1971. *Readers Digest*, 50(12):86.
Archambeau, C.B., 1968. *Rev. Geophys.*, 6(3):241.
Aulitzky, H., 1968. *Österr. Wasserwirtsch.*, 20:90.
Aulitzky, H., 1970. *Wildbach- und Lawinenverbau*, 34:31.
Bagnold, R.A., 1941. *The Physics of Blown Sand and Desert Dunes.* Methuen, London.
Bagnold, R.A., 1954. *Proc. Roy. Soc. Lond., Ser. A*, 225:49.
Ballard, G.E.H. and McGraw, R.W., 1966. *Publ. Int. Assoc. Sci. Hydrol.*, 69:160.
Barazangi, M. and Dorman, J., 1969. *Bull. Seismol. Soc. Am.*, 59(1):369.
Bardwell, G.E., 1970. *Geol. Soc. Am., Eng. Geol. Case Histories*, 8:33.
Barnes, H.H., 1967. *Roughness Characteristics of Natural Channels.* U.S. Geol. Surv. Water Supply Pap., No.1849, Washington, D.C.
Båth, M., 1953. *Gerlands Beitr. Geophys.*, 63(3):173.
Båth, M., 1956. *Sver. Geol. Unders., Årsbok, Ser. C*, 545:1.
Båth, M., 1966a. *Phys. Chem. Earth*, 7:115.
Båth, M., 1966b. *Scientia*, (7)60(4):1.
Båth, M., 1967. *Int. Dict. Geophys.*, Pergamon, Oxford, 1:424.
Båth, M., 1969. *Handbook on Earthquake Magnitude Determinations*, 2nd ed. Seismol. Inst., Uppsala.
Båth, M., 1970. *Scientia*, (7)64(6):1.
Båth, M. and Duda, S., 1964. *Ann. Geofis. (Roma)*, 17:353.
Battan, L.J., 1961. *The Nature of Violent Storms.* Doubleday-Anchor, New York, N.Y.
Baule, H., 1967. *Ber. Ver. Deut. Ing.*, 113:83.
Bazant, Z., 1966. *Proc. World Conf. Earthquake Eng., 3rd, Wellington*, 1:1.
Beaty, C.B., 1972. *Can. Geotech. J.*, 9:219.
Behre, C., 1933. *J. Geol.*, 41:622.
Benioff, H., 1951. *Bull. Geol. Soc. Am.*, 62:331.
Benioff, H., 1955. *Geol. Soc. Am. Spec. Pap.*, 62:61.
Benioff, H., 1964. *Science*, 413:1399.
Benioff, H. and Gutenberg, B., 1951. In: B. Gutenberg (Editor), *The Internal Constitution of the Earth*. New York, Dover.
Ben-Menahem, A., 1960. *Gerlands Beitr. Geophys.*, 69:68.
Ben-Menahem, A., 1962. *J. Geophys. Res.*, 67(1):345.
Ben-Menahem, A. and Rosenman, M., 1972. *J. Geophys. Res.*, 77:3097.
Berckhemer, H., 1962. *Gerlands Beitr. Geophys.*, 71:5.
Berger, K., 1968. *Umsch. Wiss. Tech.*, 1968(15):455.
Berger, K., 1971. *Bull. Schweiz. Elektrotech. Ver.*, 62(8):397.
Berger, K. and Vogelsanger, E., 1965. *Bull. Schweiz. Elektrotech. Ver.*, 56(1):2.
Bieniawski, Z.I., Denkhaus, H.G. and Vogler, U.W., 1969. *Int. J. Rock Mech. Mining Sci.*, 6:323.
Biot, M.A., 1941. *J. Appl. Phys.*, 12(2):155.
Biot, M.A., 1955. *J. Appl. Phys.*, 26(2):182.

Birch, F., 1966. *Geol. Soc. Am. Mem.*, 97:97.
Bishop, A.W., 1959. *Teknisk Ukeblad*, 106:39.
Bishop, A.W., 1965. *Géotechnique*, 15:79.
Bishop, A.W. and Bjerrum, L., 1961. *Publ. Nor. Geotek. Inst. Oslo*, 34:1.
Bishop, A.W. and Gibson, R.E., 1964. *Nat. Res. Counc. Can. STP*, 316:435.
Bishop, A.W. and Lovenbury, H.T., 1969. *Proc. Int. Conf. Soil Mech. Found. Eng., 7th, Mexico City*, 1:29.
Bjerrum, L., 1954. *Publ. Nor. Geotek. Inst.*, 5:1.
Bjerrum, L., 1967. *Géotechnique*, 17(2):82.
Bjerrum, L., 1968. *Schweiz. Bauztg.*, 86(31):1.
Bjerrum, L., 1971a. *Publ. Nor. Geotek. Inst.*, 88(3):1.
Bjerrum, L., 1971b. *Publ. Nor. Geotek. Inst.*, 89:1.
Bjerrum, L. and Jørstad, A.F., 1968. *Publ. Nor. Geotek. Inst.*, 79:1.
Bjerrum, L. and Lo, K.Y., 1963. *Géotechnique*, 13:147.
Bjerrum, L., Kringstad, S. and Kummeneje, O., 1961. *Proc. Int. Conf. Soil Mech., 5th, Paris*, 1:29.
Boldizsár, T., 1970. *Pure Appl. Geophys.*, 80:260.
Bolt, B.A., 1973. *Proc. World Conf. Earthquake Eng., 5th*, Pap. No.242.
Bonilla, M.G., 1970. *U.S. Geol. Surv. Prof. Pap.*, 545D:D1.
Borcherdt, R.D., 1970. *Bull. Seismol. Soc. Am.*, 60(1):29.
Borowicka, H., 1968. *Mitt. Inst. Grundbau Bodenmech. T.H. Wien*, 10:1.
Bossolasco, M., Dagnino, I., Elena, A. and Flocchini, H., 1972. *Studi in onore Giuseppina Aliverti*. Univ. di Napoli, Napoli, p.213.
Bottinga, Y. and Weill, D.F., 1972. *Am. J. Sci.*, 272:438.
Bouchon, M., 1973. *Bull. Seismol. Soc. Am.*, 63:615.
Bowen, A.J., 1969. *J. Geophys. Res.*, 74:5467.
Brace, W.F. and Martin, R.J., 1968. *Int. J. Rock Mech. Min. Sci.*, 5:415.
Bradley, W.C., Fahnestock, R.K. and Rowekamp, E.T., 1972. *Bull. Geol. Soc. Am.*, 83:1261.
Brady, A.G. and Hudson, D.E., 1973. *Proc. World Conf. Earthquake Eng., 5th, Rome*. Pap. No.154.
Braslau, D. and Lieber, P., 1968. *Bull. Seismol. Soc. Am.*, 58:613.
Brockamp, B., Lliboutry, L. and Muller, L., 1967. *Gletscherabbruch Mattmark, Sachverständigen-Gutachten über die Eislawine vom 30. August 1965, erstattet an das Instruktionsgericht Visp*. Public Reports of the Court Proceedings, Visp.
Brooks, E.M., 1951. In: *Compendium of Meteorology*. Am. Meteorol. Soc., Washington, D.C., p.673.
Brown, C.B., Evans, R.J. and LaChapelle, E.R., 1972. *J. Geophys. Res.*, 77(24):4570.
Brückl, E., 1972. *J. Glaciol.*, 11(62):215.
Brückl, E. and Scheidegger, A.E., 1972. *Rock Mech.*, 4:237.
Brückl, E. and Scheidegger, A.E., 1973. *Géotechnique*, 23:101.
Brunner, F. and Scheidegger, A.E., 1973. *Rock Mech.*, 5:43.
Bullen, K.E., 1965. *An Introduction to the Theory of Seismology*. Cambridge Univ. Press, London.
Bukovskaya, E.M. and Kuznetsova, K.I., 1971. *Bull. (Izv.) Acad. Sci. USSR, Earth Phys.*, 1971(2):11.
Byerlee, J.D. and Brace, W.F., 1972. *Bull. Seismol. Soc. Am.*, 62(2):657.

Byerly, P. and Stauder, W., 1958. *Earthquake Notes*, 29(3):17.
Byers, H.R. and Braham, R.R., 1949. *Report of the Thunderstorm Project.* U.S. Government Printing Office, Washington, D.C.
Callander, R.A., 1969. *J. Fluid Mech.*, 36:465.
Caloi, P., 1970. *Ann. Geofis. (Roma)*, 23(4):281.
Caloi, P., 1972. *Ann. Geofis. (Roma)*, 25:105.
Caloi, P. and Migani, M., 1972. *Ann. Geofis. (Roma)*, 25(1):15.
Caloi, P. and Spadea, M.C., 1958. *Ann. Geofis. (Roma)*, 11:191.
Caloi, P. and Spadea, M.C., 1959. *Ann. Geofis. (Roma)*, 12:489.
Campbell, W.J. and Rasmussen, L.A., 1969. *Can. J. Earth Sci.*, 6:979.
Caputo, M., 1961. *J. Geophys. Res.*, 66:1479.
Caputo, M., Pieri, L. and Rossi Tesi, F., 1972. *Ann. Geofis. (Roma)*, 25(1):55.
Carder, D.S., 1970. *Geol. Soc. Am. Eng. Geol. Case Histories*, 8:51.
Carrigy, M.A., 1970. *Sedimentology*, 14:147.
Casagrande, A. and Rivard, P.J., 1959. *Publ. Norg. Geotek. Inst.*, 31:5.
Chagnon, S.A., 1964. *Mon. Weather Rev.*, 1964(3):115.
Chandler, R.J., 1973. *J. Geol.*, 81:1.
Chervet, A., 1971. *Schweiz. Bauztg.*, 89(24).
Chervet, A. and Dallèves, P., 1970. *Schweiz. Bauztg.*, 88(19).
Chinnery, M.A., 1960. *J. Geophys. Res.*, 65(11):3852.
Chinnery, M.A., 1964. *J. Geophys. Res.*, 69:2085.
Chinnery, M.A., 1969a. *Publ. Dominion Obs. Ottawa*, 37(7):211.
Chinnery, M.A., 1969b. *Bull. Seismol. Soc. Am.*, 59(5):1969.
Chopra, A.K. and Chakrabarti, P., 1973. *Bull. Seismol. Soc. Am.*, 63:381.
Chow, V.T., 1964. *Handbook of Applied Hydrolgy.* McGraw-Hill, New York, N.Y.
Clar, E. and Weiss, P., 1965. *Berg Hüttenm. Leoben*, 110(12):447.
Clarke, H.H., 1962. *Div. Meteorol. Phys., CSIRO, Melbourne, Tech. Pap.*, No.13.
Cleaves, A.B., 1961. *Landslide Investigations, U.S. Dept. Public Roads Handbook.* U.S. Government Printing Office, Washington, D.C.
Collins, I., 1968. *J. Glaciol.*, 7:199.
Committee on Strong Motion Earthquake Observation, Tokyo, 1973. *Strong Motion Earthquake Records in Japan.* Earthquake Research Inst., Univ. of Tokyo, Tokyo, 15 Vols. (to 1973).
Comniakis, P., Drakopoulos, J., Moumoulichis, G. and Papazachos, B., 1968. *Ann. Geofis. (Roma)*, 21(1):39.
Constantinescu, L., 1965. *Rev. Roum. Géol., Géophys., Géogr., Sér. Géophys.*, 9(1):3.
Cook, E.F., 1966. *Earth Sci. Rev.*, 1:155.
Crandell, D. and Varnes, D., 1960. *Bull. Geol. Soc. Am.*, 71:1846.
Cressman, G.P., 1969. *Bull. Am. Meteorol. Soc.*, 50(11):850.
Culling, W.E.H., 1963. *J. Geol.*, 71:127.
Cunningham, M.J., 1972. *J. Hydrol. N.Z.*, 11(1):47.
Daneš, Z.F., 1972. *J. Geophys. Res.*, 77(8):1430.
Davis, R.A. and Fox, W.T., 1972. *J. Geol.*, 80:484.
Dawson, G.A. and Warrender, R.A., 1973. *J. Geophys. Res.*, 78(18): 3619.
De Bremaecker, J.C., 1959. *J. Geophys. Res.*, 64(11):1961.
De Crécy, L., 1966. *Publ. Int. Assoc. Hydrol. Sci.*, 69:369.
Defant, A., 1957. *Encycl. Phys. (Springer)*, 48:846.
De Quervain, M.R., 1966. *Publ. Assoc. Int. Sci. Hydrol.*, 69:410.

De St. Venant, B., 1871. *C.R. Acad. Sci., Paris,* 73.
De Wiest, R.J.M., 1965. *Geohydrology.* Wiley, New York, N.Y.
Dirby, W., 1969. *Water Resour. Res.,* 5(4):778.
Donn, W.L. and Wolf, D.M., 1972. *Limnol. Oceanogr.,* 17(4):639.
Dracos, T., 1970. *Schweiz. Bauztg.,* 88(19).
Duckstein, L., Fogel, M.M. and Kisiel, C.C., 1972. *Water Resour. Res.,* 8(2):410.
Duda, S.J., 1970. *Bull. Seismol. Soc. Am.,* 60:1479.
Du Montelle, P.B., Hester, N.C. and Cole, R.E., 1971. *Ill. State Geol. Surv., Environ. Geol. Notes,* 48:1.
Dzhibladze, E.A., 1971a. *Bull. (Izv.) Acad. Sci. USSR, Earth Phys.,* 1971(5): 89.
Dzhibladze, E.A. 1971b. *Bull. (Izv.) Acad. Sci. USSR, Earth Phys.,* 1971(9): 72.
Eagleson, P.S., 1972. *Water Resour. Res.,* 8(4):878.
Easton, A.K., 1972. *Can. J. Earth Sci.,* 9:857.
Eaton, G., 1964. *J. Geol.,* 72(1):1.
Eide, O. and Bjerrum, L., 1955. *Géotechnique,* 5(1).
Emery, J.J. and Finn, W.D.L., 1972. *Creep Problems in Soil Mechanics.* Paper at the Symp. Appl. Solid Mech., Univ. Waterloo, Ont.
Engelund, F., 1970. *J. Fluid Mech.,* 42:125.
Essa, 1966. *ESSA Symposium on Earthquake Prediction.* Government Printing Office., Washington, D.C.
Evans, D.M., 1970. *Geol. Soc. Am., Eng. Geol. Case Histories,* 8:25.
Evison, F.F., 1966a. *Bull. Seismol Soc. Am.,* 57:9.
Evison, F.F., 1966b. *Nature,* 211:273.
Evison, F.F., 1967a. *Vesiac Rept.,* 7885-1-X:173.
Evison, F.F., 1967b. *N.Z., J. Geol. Geophys.,* 10(2): 479.
Fara, H.D., 1964. *Bull. Seismol Soc. Am.,* 54: 1491.
Farran, J. and Thénoz, B., 1965. *L'alitérabilité des roches, les fractures, sa prévision.* Ann. de l'I.T.B.T.P., No. 215.
Fellenius, W., 1927. *Erdstatische Berechnungen.* Ernst, Berlin.
Fellenius, W., 1948. *Erdstatische Berechnungen.* Neue Aufl., Ernst, Berlin.
Ferraes, S.G., 1967. *J. Geophys. Res.,* 72:3741.
Finn, W.D.L. and Emery, J.J., 1970. *Deformations in creeping slopes.* Discussion prepared for the 2nd Congress of the International Society of Rock Mechanics, Belgrade.
Finn, W.D.L. and Emery, J.J., 1972. *Proc. Int. Geol. Congr., 24th, Montreal,* 13:117.
Finn, W.D.L. and Miller, R.I.S., 1973. *Proc. World Conf. Earthquake Eng., 5th, Rome,* Pap.42.
Finn, W.D.L., Bransby, P.L. and Pickering, D.J., 1970. *Proc. Am. Soc. Civ. Eng., J. Soil Mech. Found. Div.,* 96(SM6)1917.
Fisher, R.V., 1964. *J. Geophys. Res.,* 69(2):341.
Fisher, R.V., 1966. *Am. J. Sci.,* 264(5):350.
Fleisher, P.J. and Sales, J.K., 1972. *Bull. Geol. Soc. Am.,* 83(3):905.
Fletcher, N.H., 1970. *The Chemical Physics of Ice.* Cambridge Univ. Press, London.
Flora, S.D., 1953. *Tornadoes of the United States.* Univ. Oklahoma Press, Norman, Okla.

Forel, F.A., 1884. *Arch. Sci. Phys. Nat.*, 11:147.
Fox, W.T. and Davis, R.A., 1973. *Bull. Geol. Soc. Am.*, 84:1769.
Frenkel, Y.I., 1947. *J. Franklin Inst.*, 243:287.
Frütiger, H., 1966. *Publ. Int. Assoc. Sci. Hydrol.*, 69:243.
Fuchs, A., 1949. *Veröff. Mus. Ferdinandeum Innsbruck*, 26:101.
Fujita, T., 1959. *U.S. Weather Bur. Contract No. Cwb 9530.* Rept. No.5 University of Chicago, Chicago, Ill.
Fujita, T., 1971. *Satellite and Mesometeorology Research Project, Res. Pap.* 91, Univ. Chicago, Chicago, Ill.
Fujita, T., Bradbury, D.L. and Van Thullenar, C.F., 1970. *Mon. Weather Rev.*, 98(1):29.
Fuquay, D.M., 1967. *Ground Level Climatology.* Am. Assoc. Advan. Sci., Washington, D.C.
Fuquay, D.M., Taylor, A.R., Hawe, R.G. and Schmid, C.W., 1972. *J. Geophys. Res.*, 77(12):2156.
Fürlinger, W.L., 1972. *Geol. Rundsch.*, 61(3).
Gombolati, G., 1972a. *IBM J. Res. Dev.*, 16(2):130.
Gombolati, G., 1972b. *Bull. Int. Assoc. Sci. Hydrol.*, 17(2):219.
Gombolati, G. and Freeze, A., 1973. *Water Resour. Res.*, 9(3):721.
Gangl, G., 1969. *Publ. Zentr. Anst. Meteorol. Geodyn. Wien*, 193:31.
Garlanger, J.E., 1972. *Géotechnique*, 22(1):71.
Gassmann, F., 1925. *Jahresber. Schweiz. Erdbebendienst*, 1925(9).
Gayskiy, V.N., 1967. *Bull. (Izv.) Acad. Sci. USSR, Earth Phys..*, 1967(7):20.
Gerber, E. and Scheidegger, A.E., 1965. *Rock Mech. Eng. Geol. Suppl.*, 2:80.
Gerber, E. and Scheidegger, A.E., 1966. *Geogr. Helv.*, 21:20.
Gerber, E. and Scheidegger, A.E., 1969. *Eclogae Geol. Helv.*, 62:401.
Gerber, E. and Scheidegger, A.E., 1973. *Z. Geomorphol., Suppl.*, 18:38.
Gerber, E. and Scheidegger, A.E., 1974. *Rock Mechanics*, 6:25.
Gibson, R.E. and Morgenstern, N., 1962. *Géotechnique*, 12:212.
Gilbert, G.K., 1904. *Bull. Geol. Soc. Am.*, 15:29.
Glasstone, S., Laidler, K. and Eyring, H., 1971. *The Theory of Rate Processes.* McGraw-Hill, New York, N.Y.,
Glen, J.W., 1955. *Proc. R. Soc., Lond., Ser. A*, 228: 519.
Gold, L.W., 1970. *Can. Geotéch. J.*, 7(4):405.
Golde, R.H., 1967. *J. Franklin Inst.*, 283:451.
Gol'din, B.M. and Lyubashevsky, S.L., 1966. *Sov. Hydrol.*, 2:179.
Goldstein, M. and Ter-Stepanian, G., 1957. *Proc. Int. Conf. Soil Mech. Found. Eng., 4th, London*, 2: 311.
Goodman, R.E. and Blake, W., 1965. *Rock Mech. Eng. Geol., Suppl.*, 2: 88.
Gopwani, M.V. and Scheidegger, A.E., 1971. *Ann. Geofis. (Roma)*, 24(1):1.
Gorshkov, G.S., 1959. *Bull. Volcanol.* (2), 20:77.
Gough, D.I. and Gough, W.I., 1970. *(i) Load-induced earthquakes at Lake Kariba. (ii) Stress and deflection in the lithosphere near Lake Kariba.* Geophys. Rept. Dept. Physics, University of Alberta, Edmonton, Alta.
Gould, J.P., 1960. *Proc. ASCE Res. Conf. Shear Strength of Cohesive Soils, Boulder, Colo.*, 115.
Gow, A.J. and Williamson, T.C., 1972. *J. Geophys. Res.*, 77(32):6348.
Gray, H., 1945. *Trans. Am. Soc. Civil Eng.*, 110:1327.
Green, J. and Short, N.M., 1971. *Volcanic Landforms and Surface Features.* Springer, Berlin.

Gubin, I.E., 1970. *Bull. (Izv.) Acad. Sci. USSR, Earth Phys.*, 1970(6):3.
Gumbel, E.J., 1941. *Ann. Math. Statist.*, 12(2):163.
Gumbel, E.J., 1943. *Trans. Am. Geophys. Union*, 24:699.
Gutenberg, B. and Richter, C.F., 1942. *Bull. Seismol. Soc. Am.*, 32:163.
Gutenberg, B. and Richter, C.F., 1949. *Seismicity of the Earth and Associated Phenomena.* Princeton Univ. Press, Princeton.
Gutenberg, B. and Richter, C.F., 1954. *Seismicity of the Earth and Associated Phenomena,* 2nd ed. Princeton Univ. Press, Princeton N.J.
Gutenberg, B. and Richter, C.F., 1956. *Bull. Seismol. Soc. Am.*, 46:105.
Haefeli, R., 1963. *Ice and Snow,* 1963:560.
Haefeli, R., 1966a. *Proc. Int. Conf. Soil. Mech.*, 6th, 134.
Haefeli, R., 1966b. *Publ. Int. Assoc. Sci. Hydrol.*, 69:316.
Haefeli, R., 1967. *Schweiz. Bauztg.*, 85(1,2):3.
Haefeli, R., Schaerer, C. and Amberg, G., 1953. *Proc. Int. Conf. Soil Mech. Found. Eng., 3rd, Zürich,* 2:175.
Hahn, P.H., 1969. *Bull. Geol. Soc. Am.*, 80:2123.
Hall, M.J., 1972. *Water Water Eng.*, 76(914):125.
Handin, J., 1969. *J. Geophys. Res.*, 74:5343.
Handin, J., Heard, H.C. and Magouirk, J.N., 1967. *J. Geophys. Res.*, 72:611.
Haque, S.M.A., 1952. *Q. J. R. Meteorol. Soc.*, 78:394.
Harrison, W.D., 1972. *J. Glaciol.*, 11(63):455.
Hastie, L.M. and Savage, J.C., 1970. *Bull. Seismol. Soc. Am.*, 60(4):1389.
Hawkes, I. and Mellor, M., 1972. *J. Glaciol.*, 11(61):103.
Healy, J., 1954. *Tangiwai Railway Disaster. Report of Board of Inquiry.* Government Printer, Wellington.
Healy, J., 1963. *Bull. Volcanol.*, 26:141.
Healy, J. and Press, F., 1959. *Bull. Seismol. Soc. Am.*, 49:193.
Healy, J. and Press, F., 1960. *Geophysics,* 25:987.
Healy, J.H., Rubey, W.W., Griggs, D.T. and Raleigh, C.B., 1968. *Science,* 161:1301.
Hédervári, P., 1963. *Bul. Volcanol.*, 25:373.
Hédervári, P., 1971. *Z. Geophys.*, 37:137.
Hédervári, P., 1972. *Ann. Geofis. (Roma),* 25(3):451.
Heezen, B.C. and Ewing, M., 1952. *Am. J. Sci.*, 250:849.
Heim, A., 1932. *Bergsturz und Menschenleben.* Fretz und Wasmuth, Zürich.
Henderson, F.M. and Wooding, R.A., 1964. *J. Geophys. Res.*, 69(8):1521.
Henkel, D.J., 1970. *Géotechnique,* 20(1):75.
Hill, R., 1950. *The Mathematical Theory of Plasticity.* Clarendon Press, Oxford.
Hill, R.D., 1971. *J. Geophys. Res.*, 76:637.
Hofmann, H., 1972. *Schematische Modellstudien zum Böschungsproblem in regelmässig geklüfteten Medien.* Diss. Tech. Univ., Karlsruhe.
Hofmann, L., 1969. *Wildbach-Lawinenverbau,* 33(1):108.
Hofmann, L., 1971. *Wildbach-Lawinenverbau,* 35(2):65.
Holmsen, P., 1953. *Géotechnique,* 3(5):187.
Horninger, G., 1958. *Geol. Bauw.*, 24:37.
Housner, G.W., 1967. In: S.K. Runcorn (Editor), *International Dictionary of Geophysics.* Pergamon Press, London, 1:392.

Housner, G.W., 1973. *Proc. World Conf. Earthquake Eng., 5th, Rome.* Invited paper on ground motion.
Housner, G.W. and Jennings, P.C., 1973. *Proc. World Conf. Earthquake Eng., 5th, Rome.* Paper No.203.
Howell, B.F., 1972. *Bull. Seismol. Soc. Am.*, 62:789.
Huang, Y.H., 1969. *J. Soil Mech. Div. Am. Soc. Civil Eng.*, 96(SM1):371.
Hubbert, M.K. and Rubey, W.W., 1959. *Bull. Geol. Soc. Am.*, 70:115.
Hudson, D.E., 1972. *Bull. Seismol. Soc. Am.*, 62:1765.
Hudson, D.E. and Uwadia, F.E., 1973. *Proc. World Conf. Earthq. Eng., 5th, Rome.* Paper No.78.
Hurst, H.E., 1951. *Trans. Am. Soc. Civil Eng.*, 116:770.
Hutchinson, G.E., 1957. *A Treatise on Limnology.* Wiley, New York, N.Y.
Hutchinson, J.N., 1965. *Publ. Nor. Geotek. Inst.*, 61:1.
Hutchinson, J.N., 1967. *Proc. Geotech. Conf. Oslo,* 1:113.
Hutchinson, J.N., 1969. *Géotechnique,* 19(1):6.
Hutchinson, J.N., 1970. *Géotechnique,* 20(4):412.
Hutchinson, J.N., 1971. *Proc. Int. Conf. Nat. Slopes Stability Conservation, Consenza.*
Hutchinson, J.N. and Bhandari, R.K., 1971. *Géotechnique,* 21(4):353.
Hwang, L.S., Butler, H.L. and Divoky, D.J., 1972. *Bull. Seismol. Soc. Am.*, 62(6):1579.
Iida, K., 1959. *J. Earth Sci. Nagoya Univ.,* 7:98.
Iida, K., 1965. *J. Earth Sci. Nagoya Univ.,* 13:115.
Iida, K., Cox, D. and Pararas-Carayannis, G., 1967. *Preliminary Catalog of Tsunamis Occurring in the Pacific Ocean.* Hawaii Inst. Geophysics, Data Rept. No.5, Honolulu.
Ikegami, R., 1968. *Zisin (2),* 21(3):190.
Imbo, G., 1965. *Ann. Oss. Vesuviano,* 7(6):3.
International Association of Volcanology, 1951/67. *Catalog of Active Volcanoes of the World.* 20 vols., Naples.
Isaacson, E., 1962. *Rock Pressure in Mines.* Mining Publ. Ltd., London.
Isacks, B.L., Sykes, R. and Oliver, J., 1967. *Bull. Seismol. Soc. Am.*, 57:935.
Ishihara, K. and Yasuda, S., 1973. *Proc. World Conf. Earthquake Eng., 5th, Rome.* Paper No.38.
Jäckli, H. 1948. *Schweizer Bauztg.,* 66(37):503.
Jagschitz, J.A. and Wakefield, R.C., 1971. *Bull. Rhode Isl. Agric. Exp. Sta.,* 408:1.
Jahns, R.H., 1943. *J. Geol.,* 51:71.
Jeffreys, H., 1958. *Mont. Notes R. Astron. Soc.,* 118:14.
John, B.S. and Sugden, D.E., 1971. *Geogr. Mag.*, 43(12):853.
Johnson, H. and Smith, B.L., 1970. *The Megatectonics of Continents and Oceans.* Rutgers Univ. Press, New Brunswick, N.J.
Jonas, J. and Müller, F., 1969. *Can. J. Earth Sci.,* 6:963.
Kamb, B., 1970. *Rev. Geophys. Space Phys.,* 8(4):673.
Kamenov, B., Iliev, I., Tsvetdov, S., Avramova, E. and Simenova, G., 1971. *Proc. Conf. Natural Slopes Stability Conserv. Cosenza,* Pap. No. 5.
Kamra, A.K., 1972. *J. Geophys. Res.,* 77:5856.
Kármán, T., 1930. *Nachr. Ges. Wiss. Göttingen, Math. Phys. Kl.,* 1930:58.
Kárník, V., 1960. *Seismicity of Europe.* Reidel, Dordrecht.

Kárník, V., 1965. *Studia Geophys. Geodet.*, 9:236.
Kárník, V., 1971. *Seismicity of the European Area II.* Reidel, Dordrecht.
Kartashov, S.N., 1966. *Publ. Int. Assoc. Sci. Hydrol.*, 69:114.
Kasser, P., 1970. *Z. Gletscherk. Glazialgeol.*, 6(1,2):141.
Kennedy, J.F., 1969. *Ann. Rev. Fluid Mech.*, 1:147.
Kent, P.E., 1966. *J. Geol.*, 74:79.
Kerr, P.F. and Drew, I.M., 1967. *Calif. Div. Mines Geol., Short Contrib.*, SR 100:1.
Keulegan, G.H., 1970. *Proc. Am. Soc. Civ. Eng., J. Waterways Harb. Div.*, WW2:219.
Kieslinger, A., 1958. *Geol. Bauwes.*, 24:95.
King, C.Y. and Knopoff, L., 1968. *Bull. Seismol. Soc. Am.*, 58(1):249.
King, C.Y. and Knopoff, L., 1969. *Bull. Seismol. Soc. Am.*, 59:269.
Kirkby, M.J., 1967. *J. Geol.*, 75:359.
Kirkby, M.J., 1969. *Water Resour. Res.*, 5(4):778.
Kishida, H., 1969. *Soils Found.*, 9(1):75.
Kissin, I.G., 1971. *Bull. (Izv.) Acad. Sci. USSR, Earth Phys.*, 1971(3):39.
Kjaernsli, B. and Simons, N., 1962. *Géotechnique*, 12(2):147.
Kleinschmidt, E., 1951. *Arch. Meteorol. Geophys. Bioklimatol., Ser. A*, 4:53.
Knopoff, L., 1964. *Bull. Seismol. Soc. Am.*, 54:1871.
Knopoff, L. and Gilbert, F., 1960. *Bull. Seismol. Soc. Am.*, 50:117.
Knopoff, L. and Randall, M.J., 1970. *J. Geophys. Res.*, 75:4957.
Kobayashi, Y., 1971. *J. Phys. Earth*, 19(3):217.
Kobayashi, H. and Kagami, H., 1973. *Proc. World Conf. Earthquake Eng., 5th, Rome.* Paper No.204.
Kojan, E., Foggin, G.T. and Rice, R.M., 1972. *Proc. Int. Geol. Congr., 24th*, 13:124.
Komar, P.D., 1970. *Bull. Geol. Soc. Am.*, 81:1555.
Komar, P.D., 1971. *Bull. Geol. Soc. Am.*, 82:2643.
Komar, P.D. and Inman, D.L., 1970. *J. Geophys. Res.*, 75:5914.
Koschmieder, H., 1940. *Reichsamt Wetterd. Wiss. Abh.*, VI,3.
Kövesligethy, R. de., 1900. *Seismonomia.* Modena.
Kraijenhoff van de Leur, D.A., 1966. *Recent Trends in Hydrograph Synthesis.* Committee for Hydrological Research, T.N.O., The Hague.
Krause, C.D., White, W.C., Piper, D.J.W. and Heezen, B.C., 1970. *Bull. Geol. Soc. Am.*, 81(7):2153.
Kuenen, P.H. and Sengupta, S., 1970. *Geol. Mijnbouw*, 49(2):89.
Kurzeme, M., 1971. *Proc. Austr.–N.Z. Conf. Geomech., 1st, Melbourne*, 2:45.
Lachapelle, E., 1966. *Publ. Int. Assoc. Sci. Hydrol.*, 69:350.
Lachapelle, E. and Stillman, R.M., 1966. *Publ. Int. Assoc. Sci. Hydrol.*, 69:261.
Lacroix, A., 1904. *La montagne Pelée et ses éruptions.* Masson, Paris.
Ladanyi, B., 1972. *Can. Geotech. J.*, 9(1):63.
Ladanyi, B. and Archambault, G., 1969. *Proc. Symp. Rock Mech., 11th, Berkeley* 105.
Ladanyi, B. and Archambault, G., 1972. *Proc. Int. Geol. Congr., 24th, Montreal*, 13:249.

Lang, T.E., Brown, R.E., Lawrence, W.F. and Bradley, C.C., 1973. *J. Geophys. Res.*, 78(2):339.
Langbein, W.B. and Leopold, L.B., 1966. *U.S. Geol. Surv. Prof. Pap.*, 422H
Lanser, O., 1967. *Felsmech. Ing. Geol.*, 5(1):90.
Latham, J. and Mason. B.J., 1961. *Proc. R. Soc. Lond., Ser. A*, 260:537.
Lauffer, H., Neuhauser, E. and Schober, W., 1971. *Osterr. Ing. Z.*, 4(14):101.
Lawler, E.A., 1964. In: V.T. Chow (Editor), *Handbook of Applied Hydrology*, McGraw-Hill, New York, N.Y.
Lee, T., 1972. *Bull. Seismol. Soc. Am.*, 62(6):1597.
Lensen, G.J. and Otway, P.M., 1971. *Bull. R. Soc. N. Z.*, 9:107.
Leopold, L.B., Emmett, W.W. and Myrick, R.M., 1966. *U.S. Geol. Surv. Prof. Pap.*, 352–G.
Letzmann, J., 1921. *Sitzungsber. Naturforsch. Ges. Univ. Dorpat*, 1921:28.
Letzmann, J., 1923. *Acta Comm. Univ. Dorpat*, AVI:3.
Liebowitz, H. (Editor), 1969ff. *Fracture, an Advanced Treatise in Seven Volumes.* Academic Press, New York.
Liggett, J.A. and Woolhiser, D.A., 1967. *Proc. ASCE, J. Eng. Mech. Div.*, EM2(4).
Litwiniszyn, J., 1963. *Rock Mech. Eng. Geol.*, 1:186.
Lliboutry, L., 1964a. *C.R. Acad. Sci. Paris*, 258:1577.
Lliboutry, L., 1964b. *Nature*, 202(4927):77.
Lliboutry, L., 1969. *Can. J. Earth Sci.*, 6:943.
Lliboutry, L., 1970. *Houille Blanche*, Vol. spec., 5:489.
Lliboutry, L., 1971. *J. Glaciol.*, 10(58):15.
Lo, K.Y. and Lee, C.F., 1973. *Géotechnique*, 23:1.
Lofgren, B.E., 1969. *Rev. Eng. Geol., Geol. Soc. Am.*, 2:271.
Lohnes, R.A., and Handy, R.I., 1968. *Z. Geol.*, 76:247.
Lomnitz, C., 1956. *J. Geol.*, 64:473.
Lomnitz, C., 1966a. *Bull. Seismol. Soc. Am.*, 56:247.
Lomnitz, C., 1966b. *Rev. Geophysics*, 4(3):377.
Lomnitz, C., 1967. *IBM Symp. Sci. Comp. Environ. Sci.*: 129.
Londe, P., 1968. *Ann. Inst. Tech. Bâtim. Publ.*, 21:1617.
Londe, P., Vigier, G. and Vormeringer, R., 1970. *Proc. Am. Soc. Civ. Eng., J. Soil Mech. Found. Div.*, SM4:1411.
Longuet-Higgins, M.S., 1970. *J. Geophys. Res.*, 75:6778.
Lucretius, M.A., 60 B.C. *De Rerum Natura.* Book 6.
Lukk, A.A., 1968. *Bull. (Izv.) Acad. Sci. USSR, Earth Physics*, 1968(5):83.
Lukk, A.A. and Ponomarev, 1972. *Bull. (Izv.) Acad. Sci. USSR, Earth Phys.*, 1972(5):3.
Lursmanashvili, O.V., 1973. *Bull. (Izv.) Acad. Sci. USSR, Earth Phys.*, 2:80.
Lutton. R.J., 1971. *Int. J. Rock Mech. Min. Sci.*, 8(2):143.
Mandelbrot, B., 1965. *C.R. Acad. Sci. Paris*, 260:3274.
Mandelbrot, B. and Wallis, J.R., 1968. *Water Resour. Res.*, 4(5):909.
Mandelbrot, B. and Wallis, J.R., 1969. *Water Resour. Res.*, 5(1):228.
Marinatos, S. and Imboden, O., 1972. *Natl. Geogr.*, 141:702.
Markgraf, H., 1961. *Arch. Meteorol. Geophys. Biokl.*, A12:339.
Marnyama, T., 1963. *Bull. Earthquake Res. Inst.*, 41:467.
Marsal, D. and Philipp, W., 1959. *Bull. Geol. Inst. Univ. Upps., N.S.*, 2(7):59.
Mason, B.J., 1972. *Proc. R. Soc. Lond.*, A327:433.

Matalas, N.C., 1967. *Water Resour. Res.*, 3(3):817.
Matalas, N.C. and Wallis, J.R., 1973. *Water Resour. Res.*, 9(2):281.
Matthes, F.E., 1930. *U.S. Geol. Surv. Prof. Pap.*, 160:1.
Mauk, F.J. and Johnston, M.J.S., 1973. *J. Geophys. Res.*, 78(17):3356.
McEchron, K.B., 1939. *J. Franklin Inst.*, 227:149.
McGinnis, L.D., 1964. *Circ. Ill. State Geol. Surv.*, 344:1.
Medvedev, S., Sponheuer, W. and Karnik, V., 1967. *Veröff. Inst. Geodyn. Jena*, 48:1.
Meier, F. and Post, A., 1969. *Can. J. Earth Sci.*, 6:807.
Melaragno. M.G., 1968. *Tornado Forces and their Effects on Buildings.* State Univ. Press, Manhattan, Kansas.
Melchior, P., 1966. *The Earth Tides.* Pergamon Press, London.
Menard, H.W., 1969. *J. Geophys. Res.*, 74:4827.
Mercalli, G., 1902. *Bull. Soc. Sismol. Ital.*, 8:184.
Mesri, G., 1973. *Water Resour. Res.*, 9(4):1090.
Michel, B. and Ramseier, R.O., 1969. *Rept. Nor. 5–15, Dept. Génie Civil, Univ. Laval, Québec.*
Milliman, J.D. and Emery, K.O., 1968. *Science,* 162(8):1121.
Milne, W.G., 1967. *Can. J. Earth Sci.*, 4(5):797.
Mitchell, R.J. and Eden, W.J., 1972. *Can. J. Earth Sci.*, 9:1001.
Miyamura, S., 1962. *Proc. Jap. Acad.*, 38(1):27.
Mogi, K., 1958. *Bull. Earthquake Res. Inst.*, 36:99.
Mogi, K., 1962. *Bull. Earthquake Res. Inst. Tokyo,* 40:125.
Mogi, K., 1967. *J. Geophys. Res.*, 72:5117.
Mohr, O., 1928. *Abhandlungen aus dem Gebiete der technischen Mechanik.* 3. Aufl. Ernst, Berlin.
Molnar, P., and Aggarwal, P.N., 1971. *Bull. Seismol. Soc. Am.*, 61(1):195.
Molnar, P., Jacob, K.H. and McCamy, K., 1973. *Bull. Seismol. Soc. Am.*, 63(1):101.
Moos, A.V., 1953. *Proc. Int. Conf. Soil Mech. Found. Eng., Zürich,* 3:252.
Morales, B., 1966. *Publ. Int. Assoc. Sci. Hydrol.*, 69:304.
Morton, D.M., 1971. *U.S. Geol. Surv. Prof. Pap.*, 733:99.
Moser, M., 1971. *Intraprevent,* 1971:35.
Muffler, L.J.P., White, D.E. and Truesdell, A.H., 1971. *Bull. Geol. Soc. Am.*, 82:723.
Mühleisen, R., 1957. *Encycl. Phys.* Springer, Berlin, 48:541.
Müller, A., Gyr, A. and Dracos, T., 1971. *J. Hydraul. Res.*, 9(3):373.
Müller, L., 1963. *Der Felsbau.* Enke, Stuttgart.
Müller, L., 1968. *Rock Mech. Eng. Geol.*, 6:1.
Müller, L., 1969. *Z. Dtsch. Geol. Ges.*, 119:65.
Müller, L. and Malina, H., 1968. *Felsmech. Ingenieurgeol.*, 6:216.
Murase, T., 1962. *J. Fac. Sci. Hokkaido Univ., Ser. 7,* 1(6):487.
Murrell, S.A.F., 1967. *Geophys. J. R. Astronom. Soc.*, 14:81.
Myachkin, V.I. and Zubkov, S.I., 1973. *Bull. (Izv.) Acad. Sci. USSR, Earth Phys.*, 1973(6):28.
Nakaya, U., 1954. *Snow Crystals: Natural and Artificial.* Harvard Univ. Press, Cambridge, Mass.
Nemčok, A., Pašek, J. and Rybář, J., 1972. *Rock Mech.*, 4:71.
Nemec, J., 1972. *Engineering Hydrology.* McGraw-Hill, London.

Neumann van Padang, M., 1960. *Bull. Volcanol.*, 23:181.
Newmark, N.M., 1965. *Géotechnique*, 15:139.
Newmark, N.M. and Rosenblueth, E., 1971. *Fundamentals of Farthquake Engineering* Prentice-Hall, Englewood Cliffs N.J.
Newton, J.G. and Hyde, L.W., 1971. *Alabama Geol. Surv. Circ.*, 68:1.
Nielsen, L.E., 1968. *Bull. Geol. Soc. Am.*, 79:1195.
Nielsen, L.E., 1969. *Can. J. Earth Sci.*, 6:955.
Nikolayev, N.I., 1972. *Bull. (Izv.) Acad. Sci. USSR, Earth Phys.*, 1972(8):113.
Nonveiller, E., 1965. *Proc. Int. Conf. Soil Mech. Found. Eng., 6th, Montréal*, 2:522.
Novikov, B.I., 1970. *Sov. Hydrol. Selected Pap.*, 2:124.
Nur, A., 1972. *Bull. Seismol. Soc. Am.*, 62(5):1217.
Nur, A., 1973. *Abstr. Eur. Geophys. Soc. Meet.*, 1:76.
Nur, A. and Booker, J.R., 1972. *Science*, 175:885.
Nye, J.F., 1951. *Proc. R. Soc. Lond., Ser. A*, 207:554.
Nye, J.F., 1970. *Proc. R. Soc. Lond., Ser. A*, 315:381.
Nye, J.F. and Mae, S., 1972. *J. Glaciol.*, 11:81.
Nyland, E., 1971. *Can. J. Earth Sci.*, 8:1184.
Oliver, J.E., 1966. *ESSA Symp. Earthquake Prediction*, 16.
Oliveri del Castillo, A. and Quagliarello, M.T., 1969. *Atti Conv. Assoc. Geofis. Ital.*, 18:1.
Oliveri del Castillo, A. and Quagliarello, M.T., 1970. *Boll. Soc. Nat. Napoli*, 79:49.
Omori, F., 1894. *J. Coll. Sci. Imp. Univ. Tokyo*, 7:111.
Orowan, E., 1960. *Geol. Soc. Am. Mem.*, 79:323.
Orowan, E., 1966. *Rev. Geophys.*, 4:395.
Orowan, E., 1967. *Vesiac Rept.*, 7885-1-X:275.
Orville, R.E., 1968. *J. Atmos. Sci.*, 25:827.
Orville, R.E. and Berger, K., 1973. *J. Geophys. Res.*, 78(21):4520.
Page, R., 1968. *J. Geophys. Res.*, 73:3897.
Palmer, A.C. and Nye, J.F., 1972. *J. Glaciol.*, 11(61):65.
Parlange, J.Y., 1973. *J. Hydrol.*, 18:119.
Parsons, R.C. and Hedley, D.G.F., 1966. *Int. J. Rock Mech. Min. Sci.*, 3:325.
Paterson. W.S.B., 1969. *The Physics of Glaciers*. Pergamon Press, London.
Patton, B., 1956. *J. Soil Water Conserv.*, 11:28.
Patton, F.D., 1966. *Proc. Congr. Int. Soc. Rock Mech., 1st, Lisboa*, 2:201.
Peacock, W.H. and Seed, H.B., 1968. *Proc. Am. Soc. Civ. Eng., J. Soil Mech. Found. Div.*, 94(SM3):689.
Penck, A., 1929. *Die geomorphologische Analyse*. J. Engelhorns Nachf., Stuttgart.
Peng, S. and Podnieks, E.R., 1972. *Int. J. Rock Mech. Min. Sci.*, 9:699.
Perla, R.I., 1972. *J. Glaciol.*, 11(63):447.
Perla, R.I. and Lachapelle, E.R., 1970. *J. Geophys. Res.*, 75(36):7619.
Peterschmitt, E., 1952. *Publ. Bull. Centr. Int. Seismol. Trav. Sci.*, A18:183.
Petit, J.R., 1971. *Bilans annuels, transformation du névé en glace, teneur en eau de la glace profonde*, Diss., Univ. Grenoble, Grenoble.
Petterson, E.K., 1916. *Tekn. Tidskr.* 46:289.

Petzny, H., 1967. *Felsmech. Ingenieurgeol., Suppl.* 3:37.
Pfeffer, R.L., 1964. *Trans. N.Y. Acad. Sci.*, (2), 26(8):984.
Pfluke, J.H. and Howell, B.F., 1964. *J. Geophys. Res.*, 69(4):653.
Pinder, G.F. and Sauer, S.P., 1971. *Water Resour. Res.*, 7(1):63.
Poland, J.F., and Davis, J.H., 1969. *Rev. Eng. Geol., Geol. Soc. Am.*, 2:187.
Post, A., 1960. *J. Geophys. Res.*, 65:3703.
Post, A., 1969. *J. Glaciol.*, 8(53):229.
Post, A., 1972. *J. Glaciol.*, 11(62):219.
Powell, R.W., 1943. *Civil Eng.*, 13:105.
Powell, W.J. and LaMoreaux, P.E., 1969. *Alabama Geol. Surv. Circ.*, 56:1.
Preisendorfer, R.W., 1972. *Surface-Wave Transport in Nonuniform Canals.* Honolulu, Hawaii Inst. Geophysics Rept. NOAA-JTRE-80.
Press, F., 1958. *Publ. Dom. Obs. Ottawa*, 20(2):271.
Press, F., 1967a. *Vesiac Rept.*, 7885-1-X:269.
Press, F., 1967b. *Vesiac Rept.*, 7885-1-X:155.
Press, F. and Brace, W.F., 1966. *Science*, 152(3729):1575.
Prior, D., Stephens, N. and Douglas, G., 1970. *Z. Geomorphol.*, 14(3):275.
Prior, D., Stephens, N. and Douglas, G., 1971. *Slopes Form and Process, Inst. of Brit. Geogr., Spec. Publ.*, 3:129.
Prokopovich, N.P., 1969. *Rock Mech.*, 1:134.
Prokopovich, N.P., 1971. *Proc. Asian Regional Conf. Soil Mech. Found. Eng., 4th, Bangkok*, 1:67.
Pühringer, A., 1961. *Arch. Meteorol. Geophys. Bioklimatol.*, A12(2):262.
Pühringer, A., 1963. *Arch. Meteorol. Geophys. Bioklimatol.*, A12(3):450.
Raleigh, C.B., Healy, J.H. and Bredehoeft, J.S., 1972. *Am. Geophys. Union, Monogr.*, 16:1.
Ranalli, G., 1969. *Ann. Geofis. (Roma)*, 22(4):359.
Ranalli, G., 1972. *Naturwiss.* 59(11):510.
Ranalli, G. and Scheidegger, A.E., 1969. *Ann. Geofis. (Roma)*, 22(3):293.
Randall, M.J., 1964a. *Bull. Seismol. Soc. Am.*, 54(5A):1283.
Randall, M.J., 1964b. *Bull. Seismol. Soc. Am.*, 54(5A):1291.
Randall, M.J., 1966. *J. Geophys. Res.*, 71(22):2597.
Randall, M.J., 1972. *J. Geophys. Res.*, 77:969.
Randall, M.J. and Knopoff, L., 1970. *J. Geophys. Res.*, 75(26):4965.
Rankine, W.J.M., 1857. *Phil. Trans. R. Soc. Lond.*, 147(1):9.
Rapp, A., 1963. *Nachr. Akad. Wiss. Göttingen, II. Math. Phys. Kl.*, 13:195.
Reich, B.M., 1970a. *Water Resour. Res.*, 6(6):1655.
Reich, B.M., 1970b. *Proc. 1st Ann. Geomorphol. Symp. Binghampton, N.Y.*, 49.
Reimer, R.B., Clough, R.W. and Raphael, J.M., 1973. *Proc. World Conf. Earthquake Eng., 5th, Rome*, Pap. No.293.
Reynolds, O. 1885. *Phil. Mag.*, 20:469.
Richards, A.F., 1963. *J. Geophys. Res.*, 68:919.
Richards, A.F., 1965. *Nature*, 207:1382.
Richards, A.F., Hersey, J.B. and McGuinness, W.T., 1962. *Mem. Serv. Geol. Portugal*, 9:27.
Richter, C.F., 1935. *Bull. Seismol. Soc. Am.*, 25:1.
Richter, C.F., 1958. *Elementary Seismology.* Freeman, San Francisco, Calif.
Richter, C.F., 1970. *Nature*, 225(5228):170.
Riecker, R.E., 1966. *Earthquake Notes*, 37(1):3.

Riecker, R.E., 1972. *Geotimes,* 17(4):15.
Riehl, H., 1954. *Tropical Meteorology.* McGraw-Hill, New York, N.Y.
Rikitake, T., 1968. *Earth Sci. Rev.,* 4(4):145.
Rikitake, T. and Yamazaki, Y., 1967. *Proc. Jap. Acad. Sci.,* 43(6):477
Rikitake, T., Yamazaki, Y., Sawada, M., Sasai, Y., Yoshino, T., Uzawa, S. and Shimomura, T., 1967. *Bull. Earthquake Res. Inst.,* 45:395.
Rinehart, J.S., 1966. *Int. J. Fracture Mech.,* 2(3):534.
Ringwood, A.E., 1967. *Earth Planet. Sci. Lett.,* 2:255.
Ritchie, A.M., 1966. *Bull. Highway Res. Board Natl. Acad. Sci.,* 17(1114):13.
Ritsema, A.R., 1967. *Vesiac Rept.,* 788-1-X:25.
Ritsema, A.R., 1970. *Phys. Earth Planet. Interiors,* 3:503.
Rittmann, A., 1960. *Vulkane und ihre Tätigkeit.* Enke, Stuttgart.
Rittmann, A., 1961. *Atti Accad. Gioenia Sci. Nat. Catania,* (6)13(2):27.
Riznichenko, Yu.V., 1964. *Proc. (Dokl.) Acad. Sci. USSR,* 157(6).
Riznichenko, Yu.V., 1965a. *Bull. (Izv.) Acad. Sci. USSR, Earth Phys.,* 1965(10):7.
Riznichenko, Yu.V., 1965b. *Bull. (Izv.) Acad. Sci. USSR, Earth Phys.,* 1965(11):1.
Riznichenko, Yu.V., 1969. *Bull. (Izv.) Acad. Sci. USSR, Earth Phys.,* 1969(7):3.
Riznichenko, Yu.V., Bune, V.I., Zadzharova, A.I. and Seiduzova, S.S., 1969. *Bull. (Izv.) Acad. Sci. USSR, Earth Phys.,* 1969(8):3.
Robin, G. de Q., 1969. *Can. J. Earth Sci.,* 6:919.
Robin, G. de Q. and Barnes, P., 1969. *Can. J. Earth Sci.,* 6:969.
Robson, G.R., Barr, K.G. and Luna, L.C., 1968. *Nature,* 218:28.
Roch, C., 1966. *Publ. Int. Assoc. Sci. Hydrol.,* 64:182.
Rodriguez-Iturbe, I., Mejia, J.M. and Dawdy, D.R., 1972. *Water Resour. Res.,* 8(4):921.
Rogers, W.F., 1972. *Water Resour. Res.,* 8(4):973.
Rosenblueth, E., 1973. *Proc. World Conf. Earthquake Eng., 5th, Rome.* Invited paper on Analysis of Risk.
Rosenqvist, I.T., 1953. *Géotechnique,* 3(5):195.
Rossi, M.S. de, 1883. *Bull. Volcanol. Ital.,* 10:3.
Rossmann, F., 1959. *Meteorol. Rundsch.,* 12(4):107.
Rothé, J.R., 1969a. *The Seismicity of the Earth.* UNESCO, Paris.
Rothé, J.R., 1969b. *Proc. World Conf. Earthquake Eng., 4th, Santiago de Chile;* Pap. No.A1:28.
Rothé, J.R., 1970. *Tectonophysics,* 9:2.
Rothé, J.R. and Schneider, G., 1968. *Catalogue des tremblements de terre du Fossé Rhenan.* Landeserdbebendienst, Stuttgart.
Röthlisberger, H., 1969. *Can. J. Earth Sci.,* 6:863.
Rubey, W.W. and Hubbert, M.K., 1959. *Bull. Geol. Soc. Am.,* 70:167.
Rutishauser, H., 1971. *J. Glaciol.,* 10(59).
Ryan, J.A., 1972. *J. Geophys. Res.,* 77(36):7133.
Rybař, J., 1971. *Rock Mech.,* 3:139.
Saint Amand, P., 1956. *Bull. Seismol. Soc. Am.,* 46(1).
Saito, Y., 1958. *Geophys. Mag.,* 28(3):329.
Sakuma, T. and Nagata, T., 1957. *Encycl. Phys. (Springer, Berlin),* 48:982.
Salm, B., 1966. *Publ. Int. Assoc. Sci. Hydrol.,* 69:199.

Saragoni, G.R. and Hart, G.C., 1973. *Proc. World Conf. Earthquake Eng., 5th, Rome.* Pap. No.156.
Sato, R., 1969. *J. Phys. Earth,* 17(2):101.
Savage, J.C., 1969. *Bull. Seismol. Soc. Am.,* 59:381.
Savage, J.C. and Mansinha, L., 1963. *J. Geophys. Res.,* 68(23):6345.
Savarenskiy, E.F. and Kirnos, D.B., 1955. *Element seysmologiy i seysmometrii.* Gos. Izdat. Tekh.-Teoret. Lit., Moscow.
Schäffner, H.J., 1961. *Z. Geophys.,* 27:164.
Scheffer, F. and Schachtschabel, P. 1970. *Lehrbuch der Bodenkunde,* 7th ed. Enke, Stuttgart.
Scheidegger, A.E., 1961. *J. Alberta Soc. Petrol. Geol.,* 9:131.
Scheidegger, A.E., 1963a. *Principles of Geodynamics.* Springer, Berlin.
Scheidegger, A.E., 1963b. *Viertelj. Schr. Natf. Ges. Zürich,* 108(2):169.
Scheidegger, A.E., 1965. *Z. Geophys.,* 31:300.
Scheidegger, A.E., 1966. *Pure Appl. Geophys.,* 65:160.
Scheidegger, A.E., 1970a. *Theoretical Geomorphology,* 2nd. ed. Springer, Berlin.
Scheidegger, A.E., 1970b. *Rock Mech.,* 2:138.
Scheidegger, A.E., 1970c. *Ann. Geofis. (Roma),* 23:27.
Scheidegger, A.E., 1970d. *Ann. Geofis. (Roma),* 23:325.
Scheidegger, A.E., 1970e. *Water Resour. Res.,* 6(3):750.
Scheidegger, A.E., 1971. *Ann. Geofis. (Roma),* 24(3):311.
Scheidegger, A.E., 1972. *Proc. Int. Geol. Congr., 24th, Montreal,* 3:79.
Scheidegger, A.E., 1973. *Rock Mech.,* 5:231.
Scheidegger, A.E., 1974. *The Physics of Flow Through Porous Media,* 3rd ed. Univ. of Toronto Press, Toronto, Ont.
Scheidegger, A.E. and Potter, P.E., 1965. *Sedimentology,* 5:289.
Scheidegger, A.E. and Potter, P.E., 1968. *Sedimentology,* 11:163.
Scheller, E., 1970. *Geophysikalische Untersuchungen zum Problem des Taminser Bergsturzes.* Diss. E.T.H., Zürich.
Schick, R., 1972. *Geol. Rundschau,* 61:896.
Schick, R. and Ruiscetti, M., 1973. *Z. Geophys.,* 39:247.
Schick, R. and Schneider, G., 1964. *Veröff. Landeserdbebendienst Baden-Württ., Modellseismol.,* 3. Teil.
Schiffman, R.L. and Stein, J.R., 1970. *J. Soil Mech. Div. Am. Soc. Civ. Eng.,* 96(SM4):1499.
Schofield, A.N., 1971. *Conf. Natl. Slopes Stability Conserv., Cosenza.*
Scholte, J.G.J., 1962. *Bull. Seismol. Soc. Am.,* 52:711.
Scholte, J.G.J. and Ritsema, A.R., 1962. *Bull. Seismol. Soc. Am.,* 52:747.
Schumm, S.A., 1967. *Science,* 155(3762):560.
Schumm, S.A. and Khan, H.R., 1972. *Bull. Geol. Soc. Am.,* 83:1755.
Schwab, F., 1965. *Bull. Schweiz. Elektrotech. Ver.,* 17:678.
Schwartz, M.L., 1968. *J. Geol.,* 76:508.
Scorer, R.S., 1966. *Sci. J.,* 2:46.
Seed, H.B., 1968. *Proc. Am. Soc. Civ. Eng., J. Soil Mech. Found. Div.,* 94(SM5):1053.
Seed, H.B. and Lee, K.L., 1966. *Proc. Am. Soc. Civ. Eng., J. Soil Mech. Found. Div.,* 92(SM6):105.

Seed, H.B. and Peacock, W.H., 1971. *Proc. Am. Soc. Civ. Eng., J. Soil Mech. Found. Div.*, 97(SM8):1099.

Serafim, J.L., 1969. In: K.G. Stagg and O.C. Zienkiewicz (Editors), *Rock Mechanics in Engineering Practice.* Wiley, London, p.55.

Shabbar, M., 1969. *Naturaliste Can.*, 65(1):651.

Sharpe, C.F.S., 1938. *Landslides and Related Phenomena.* Columbia Univ. Press, New York, N.Y.

Shaw, H.R., 1969. *J. Petrol.*, 10:510.

Shaw, H.R., 1972. *Am. J. Sci.*, 272:870.

Shebalin, N.V., 1968. *Coll. (Sb.) Seismic Mapping of the USSR.* Nauka Press, Moscow.

Shepard, F.P., 1963. In: *Essays in Marine Geology in Honour of K.O. Emery.* Univ. Southern Calif. Press, Los Angeles, Calif. p.1.

Shepard, F.P., 1964. *Science*, 143(3606):574.

Sherman, L.K., 1932. *Eng. News.-Rec.*, 108:501.

Shibata, T. and Yukimoto, H., 1970. *Proc. Jap. Soc. Civ. Eng.*, 176:1.

Shlien, S. and Toksöz, M.N., 1970. *Bull. Seismol. Soc. Am.*, 60(6):1765.

Shoda, M., 1966. *Publ. Int. Assoc. Sci. Hydrol.*, 69:215.

Showalter, A.K., 1943. *Preliminary Report of Tornadoes.* U.S. Weather Bureau, Washington, D.C.

Showalter, A.K., 1953. *Bull. Am. Meteorol. Soc.*, 34(6):250.

Shreve, R.L., 1959. *Geology and Mechanics of the Blackhawk Landslide.* Diss. Cal. Tech., Pasadena, Calif.

Shreve, R.L., 1968a. *Geol. Soc. Am. Spec. Pap.*, 108:27.

Shreve, R.L., 1968b. *Bull. Geol. Soc. Am.*, 79(5):653.

Silius Italicus, T.C.A., ca. 90., *Punica*, 3:521.

Silver, M.L. and Seed, H.B., 1971. *Proc. Am. Soc. Civ. Eng., J. Soil Mech. Found. Div.*, 97(SM8):1081.

Simpson, G.C., 1909. *Phil. Trans. R. Soc. Lond.*, A209:379.

Simpson, G.C. and Scrase, F.J., 1937. *Proc. R. Soc. Lond.*, A161:309.

Simpson, J.F., 1968. *Earth Planet. Sci. Lett.*, 3:417.

Singer, S., 1971. *The Nature of Ball Lightning.* Plenum Press, New York, N.Y.

Skempton, A. and Hutchinson, J., 1969. *State of the Art Volume of the Int. Conf. Soil Mech. Found. Eng., 7th, Mexico.*

Smalley, I.J., 1972. *Trans. Inst. Miner. Met.*, A81:31.

Smith, W.E.T. and Milne, W.G., 1969. *Publ. Dom. Obs. Ottawa, Seismol. Ser.*, 1969-2:1.

Sobolev, G.A., Morozov, V.N. and Migunov, N.I., 1972. *Bull. (Izv.) Acad. Sci. USSR, Earth Phys.*, 1972(2):73.

Solovyev, S.L., 1959. *Publ. Bur. Centr. Seismol. Int.*, A20:39.

Spence, E.S., 1973. *Can. J. Earth Sci.*, 10:130.

Spurek, M., 1972. *Studia Geograph., Brno*, 19:1.

Srikantia, S.V. and Bhargava, O.N., 1972. *Eng. Geol.*, 6(3):191.

Stauder, W., 1962. *Adv. Geophys.*, 9:1.

Steinhauser, F., 1934. *Gerl. Beitr. Geophys.*, 6:466.

Steinhauser, P., 1972. *Österr. Ing.-Z.*, 15(3):72.

Steketee, J., 1958a. *Can. J. Phys.*, 36:192.

Steketee, J., 1958b. *Can. J. Phys.*, 36:1168.

Stiny, J., 1910. *Die Muren.* Verlag d.Wagner'schen Univ. Buchh., Innsbruck.

Stiny, J., 1941. *Geol. Bauwes.*, 13:71.

Stiny, J., 1952. *Geol. Bauwes.*, 19(1):31.
Stoker, J.J., 1957. *Water Waves.* Interscience, New York, N.Y.
Stover, C.W., 1966. *J. Geophys. Res.*, 71(10):2575.
Stroganov, A.S., 1961. *Proc. Int. Conf. Soil Mech. Found. Eng., 5th,* 2:721.
Sturgul, J.R. and Scheidegger, A.E., 1967. *Rock Mech. Eng. Geol.*, 5:137.
Sugimura, A., 1963. *Kagaku (Science)*, 33(9):489.
Sugimura, A. and Uyeda, S., 1967. *Tectonophysics,* 5(1):25.
Sutton, O.G., 1943. *Micrometeorology.* McGraw-Hill, New York, N.Y.
Sykes, L.R., 1963. *J. Geophys. Res.*, 68(21):5999.
Sykes, L.R., 1969. *Am. Geophys. Union, Geophys. Monogr.*, 13:148.
Sylvester, A.G., Smith, S.W. and Scholz, C.H., 1970. *Bull. Seismol. Soc. Am.*, 60:1047.
Sytinskiy, A.P., 1963. *Geomagn. Aeron.*, 3:120.
Tamrazyan, G.P., 1968. *Icarus,* 9:574.
Tanaka, T., Yoshizawa, S., Morishita, T., Osada, K. and Osawa, Y., 1973. *Proc. World Conf. Earthquake Eng., 5th Rome,* Pap. No.73.
Tanimoto, K. and Suga, Y., 1971. *Proc. Asian Reg. Conf. Soil Mech. Found. Eng., 4th, Bangkok,* 1:235.
Tannehill, I.R., 1952. *Hurricanes, their Nature and History.* Princeton Univ. Press, Princeton, N.J.
Taylor, D.W., 1937. *J. Boston Soc. Civ. Eng.*, 24:197.
Taylor, D.W., 1948. *Fundamentals of Soil Mechanics.* J. Wiley, New York, N.Y.
Terashima, T., 1970. *Bull. Int. Inst. Seismol. Earthquake Eng. (Tokyo),* 7:93.
Ter-Stepanian, G., 1965. *Felsmech. Ing. Geol.*, 3:43.
Terzaghi, K., 1943. *Theoretical Soil Mechanics.* Wiley, New York, N.Y.
Terzaghi, K., 1962. *Géotechnique,* 12(4):251.
Terzaghi, K. and Fröhlich, O.K., 1936. *Theorie der Setzung von Tonschichten.* Deuticke, Wien.
Terzaghi, K. and Peck, W., 1949. *Soil Mechanics in Engineering Practice.* Wiley, New York, N.Y.
Tesseyre, R., 1961. *Bull. Acad. Polon. Sci., Ser. Sci. Math. Astr. Phys.*, 9(5):423.
Tesseyre,R., 1969. *Publ. Dom. Obs. Ottawa,* 37(7):199.
Thakur, R.T. and Scheidegger, A.E., 1970. *J. Hydrol.*, 12:25.
Thom, H.C.S., 1960. *J. Geophys. Res.*, 65(1):213.
Thom, H.C.S., 1963. *Mon. Weather Rev.*, 91:730.
Thomson, S., 1970. *Can. Geotech. J.*, 7(2):157.
Tocher, D., 1958. *Bull. Seismol. Soc. Am.*, 48:147.
Tocher, D., 1973. *Proc. World Conf. Earthquake Eng., 5th, Rome.* Paper No. 160b.
Todorović, P., 1970. *Ann. Math. Stat.*, 4(3):1059.
Todorović, P. and Rousselle, J., 1971. *Water Resour. Res.*, 7(5):1144.
Todorović, P. and Woolhiser, D.A., 1972. *Water Resour. Res.*, 8(6):1433.
Todorović, P. and Zelenhasić, E., 1970. *Water Resour. Res.*, 6(6):1641.
Tokarev, P.I., 1970. *Bull. (Izv.) Acad. Sci. USSR, Earth Phys.*, 1970(3):15.
Tomashevskaya, I.S. and Kamidullin, M.Ya., 1972. *Bull. (Izv.) Acad. Sci. USSR, Earth Phys.*, 1972(3):12.

Torelli, L., 1973. *The Analysis of Monthly Hydrologic Time Series.* Ph. D. Thesis, Univ. Illinois, Urbana.

Torelli, L. and Chow, V.T., 1972. *Proc. Int. Symp. Uncertainties Hydrol. Time Ser., Tucson,* 2(11):1.

Trifunac, M.D., 1972. *Bull. Seismol. Soc. Am.,* 62:1283.

Trifunac, M.D. and Hudson, D.E., 1971. *Bull. Seismol. Soc. Am.,* 61:1393.

Tschopp, J., 1972. *Schweiz. Bauztg.,* 90(27).

Tsuboi, C., 1965. *Proc. Jap. Acad.,* 41:392.

Tsuya, H., 1955. *Bull. Earthquake Res. Inst.,* 33(3).

Twidale, C.R., 1964. *Trans. Inst. Brit. Geogr.,* 34:91.

Uman, M.A., 1969. *Lightning.* McGraw-Hill, New York, N.Y.,

Uman, M.A., McLain, D.K., Fisher, R.J. and Krider, E.P., 1973a. *J. Geophys. Res.,* 78(18):3530.

Uman, M.A., McLain, D.K., Fisher, R.J. and Krider, E.P., 1973b. *J. Geophys. Res.,* 78(18):3523.

Unesco, 1971. *The Surveillance and Prediction of Volcanic Activity.* Unesco, Paris.

United States National Weather Service, 1972. *Hurricane Agnes.* Silver Spring, Md.

Van der Knaap, W. and Eijpe, R., 1968. *Sedimentology,* 11:115.

Van Dorn, W.G., 1965. *Adv. Hydrosci.,* 2:1.

Vargas, M. and Pichler, E., 1957. *Proc. Int. Conf. Soil Mech. Found. Eng., 4th, London,* 2:394.

Varnes, D.J. and Lee, F.T., 1972. *Bull. Geol. Soc. Am.,* 83:2863.

Veitch, L.G. and Shephard, K.J., 1971. *Water Resour. Res.,* 7(6):1469.

Vere-Jones, D. and Davies, R.B., 1966. *N.Z. J. Geol. Geophys.,* 9:251.

Vere-Jones, D., Turnovsky, S. and Eiby, G.A., 1964. *N.Z. J. Geol. Geophys.,* 7:722.

Vischer, D., 1970. *Tech. Rundsch.,* 53:54.

Vischer, D., 1971. *Schweiz. Bauztg.,* 89(12).

Vischer, D. and Moser, U., 1971. *Techn. Z.,* 68(15.7.71).

Voellmy, A., 1955. *Schweiz. Bauztg.,* 73(12):159.

Voight, B., 1969. *Mem. Am. Assoc. Petrol. Geol.,* 12:955.

Volger, O., 1858. *Untersuchungen über das Phänomen der Erdbeben in der Schweiz.* Gotha.

Volkart, P., 1972. *Mitt. Versuchsanst. Wasserbau, Hydrol. Glaziol. ETH Zürich,* 6:1.

Vonnegut, B., 1960. *J. Geophys. Res.,* 65:203.

Vvedenskaya, A.V., 1958. *Bull. (Izv.) Acad. Sci. USSR, Geophys. Ser.,* 1958:175.

Vvedenskaya, A.V., 1959. *Bull. (Izv.) Acad. Sci. USSR, Geophys. Ser.,* 1959:516.

Vvedenskaya, A.V., 1960. *Bull. (Izv.) Acad. Sci. USSR, Geophys. Ser.,* 1960:513.

Vvedenskaya, A.V., 1961. *Bull. (Izv.) Acad. Sci. USSR, Geophys. Ser.,* 1961:261.

Vvedenskaya, A.V., 1965. *Bull. (Izv.) Acad. Sci. USR, Phys. Earth,* 1965(1):3.

Walsh, J.B., 1968. *J. Geophys. Res.,* 73(2):761.

Ward, P.L., 1972. *Geothermics,* 1(1):3.
Wardle, L.J. and Gerrard, C.M., 1972. *Rock Mech.,* 4:155.
Washburn, A.L., 1970. *Z. Geomorphol. Suppl.,* 9:102.
Waters, A.C., 1955. *Bull. Geol. Soc. Am.,* 66:663.
Weertman, J., 1957. *J. Glaciol.,* 3:33.
Weertman, J., 1962. *Publ. Int. Assoc. Sci. Hydrol.,* 58:31.
Weertman, J., 1964. *Bull. Seismol. Soc. Am.,* 54(4):1035.
Weertman, J., 1967. *J. Geophys. Res.,* 72(2):521.
Weertman, J., 1969. *Can. J. Earth Sci.,* 6:929.
Weertman, J., 1971a. *J. Geophys. Res.,* 76(26):6485.
Weertman, J., 1971b. *J. Geophys. Res.,* 76(35):8544.
Weertman, J., 1972. *Rev. Geophys. Space Phys.,* 10(1):287.
Wegener, A., 1917. *Wind- und Wasserhosen in Europa.* Vieweg, Braunschweig.
Wehrmann, A., 1950. *Die überströmten Querwerke der Wildbachverbauung.* C. Gerold's Sohn, Wien.
Welsch, W. and Kinzl. H., 1970. *Z. Gletscherk. Glazialgeol.,* 6:181.
White, E.L. and Reich, B.M., 1970. *J. Hydrol.,* 10:193.
White, J.W., 1973. *J. Geophys. Res.,* 78(14):2438.
Wickman, F.E., 1966. *Ark. Mineral. Geol.,* 4:291.
Wiegel, R.L., 1970. *Earthquake Engineering.* Prentice-Hall, Englewood Cliffs, N.J.
Wilcock, D.N., 1971. *Bull. Geol. Soc. Am.,* 82:2159.
Williams, G. and Guy, H., 1971. *Environmental Geomorphology.* D.R. Coates (Editor), State Universities of New York, Binghampton, N.Y.
Williams, P.J., 1957. *Am. J. Sci.,* 255:705.
Williams, P.J., 1959. *Am. J. Sci.,* 257:481.
Williams, P.J. and Moore, J.G., 1973. *Geotimes,* 18(8):14.
Wilson, C.T.F., 1920. *Phil. Trans. R. Soc. Lond.,* A221:73.
Wilson, C.T.F., 1929. *J. Franklin Inst.,* 208:1.
Wilson, J.W. and Chagnon, S.A., 1971. *Circ. Illinois State Water Surv., Urbana,* 103:1.
Witherspoon, P.A. and Maini, Y.N.T., 1973. *Proc. World Conf. Earthquake Eng., 5th, Rome.* Pap. No.297.
Wittke, W., 1964. *Felsmech. Ingenieurgeol., Suppl.,* 1:103.
Wohlenberg, J., 1968. *Veröff. Bayr. Akad. Wiss., Komm. Int. Erdmess.,* 23:1.
Wood, A., 1942. *Proc. Geol. Assoc. Lond.,* 53:128.
Wood, H.O., 1908. *Publ. Carnegie Inst. Wash.,* 87:220.
Wood, H.O. and Neumann, F., 1931. *Bull. Seismol. Soc. Am.,* 21:277.
Workman, E.J. and Reynolds, S.E., 1948. *Phys. Rev.,* 74:709.
Workman, E.J. and Reynolds, S.E., 1950. *Phys. Rev.,* 78:254.
Wu, F.T., Thomson, K.C. and Kuenzler, H., 1972. *Bull. Seismol. Soc. Am.,* 62(6):1621.
Wuerker, R.G., 1956. *Am. Inst. Miner. Metal. Eng.,* Pap. No.663-G.
Wundt, W., 1949. *Experientia,* 5(8):301.
Wyss, M. and Brune, J.N., 1968. *J. Geophys. Res.,* 73:468.
Wyss, M. and Molnar, P., 1972. *J. Geophys. Res.,* 77:1433.
Yanai, M., 1961. *J. Meteorol. Soc. Jap., Ser. II,* 39:282.
Yokoyama, I., 1956. *Bull. Earthquake Res. Inst.,* 34(2):185.
Yokoyama, I., 1957a. *Bull. Earthquake Res. Inst.,* 35(1):75.

Yokoyama, I., 1957b. *Bull. Earthquake Res. Inst.*, 35(1):99.
Yokoyama, I., 1969. *J. Phys. Earth*, 17(1):69.
Yokoyama, I., 1971. *J. Phys. Earth*, 19(3):199.
Youash, Y., 1969. *Bull. Geol. Soc. Am.*, 80:303.
Záruba, Q. and Mencl, V., 1969. *Landslides and their Control.* Elsevier, Amsterdam.
Zeller, J., 1972. *Fachveranst. 100 Jahre Hochsch. Bodenkultur in Wien*, 4:169.
Zen, M.T. and Hadikusumo, D., 1965. *Bull. Volcanol.*, 28:275.
Ziony, J.I. and Wentworth, J.M., 1973. *Proc. World Conf. Earthquake Eng., 5th, Rome.* Pap. No.210.
Zischinsky, U., 1967. *Mitt. Ges. Geol. Bergbaustud. Wien*, 17:127.
Zischinsky, U., 1969. *Rock Mech.*, 1:30.
Zitta, V.L. and Wiggert, J.M., 1971. *Water Resour. Res.*, 7(5):1341.
Zumberger, J.H. and Ayers, J.C., 1964. In: V.T. Chow (Editor), *Handbook of Applied Hydrology*. McGraw-Hill, New York, N.Y.

AUTHOR INDEX

SUBJECT INDEX

V